# DFT Study of Cu-Nitrogen Doped Graphene (Cu-N/Gr) Nanocomposite as Cathode Catalyst in Fuel Cells

*DFT Study on Cu-N/Gr Nanocomposite as Cathode Catalyst in Fuel Cells*

# DFT Study of Cu-Nitrogen Doped Graphene (Cu-N/Gr) Nanocomposite as Cathode Catalyst in Fuel Cells

***DFT Study on Cu-N/Gr Nanocomposite as Cathode Catalyst in Fuel Cells***

Roshan Thotagamuge et al.

ELIVA PRESS

ISBN: 978-99949-8-499-2



Cover Design: Eliva Press
Cover Image: Ingimage
Email: info@elivapress.com
Website: www.elivapress.com

This Eliva Press imprint is published by the registered company Eliva Press Global Ltd. part of Eliva Press S.R.L. Publishing Group

The registered company address is: Pope Hennessy Street Level 2, Hennessy Tower Port Louis, Mauritius
Eliva Press S.R.L. Publishing Group address is: Bulevardul Moscova 21, Chisinau, Moldova, Europe

# DFT Study of Cu-Nitrogen Doped Graphene (Cu-N/Gr) Nanocomposite as Cathode Catalyst in Fuel Cells

**DFT Study of Cu-Nitrogen Doped Graphene (Cu-N/Gr) Nanocomposite as Cathode Catalyst in Fuel Cells**

Yashas Balasooriya[1], Pubudu Samarasekara[2], Chee Ming Lim[3], Yuan-Fong Chou Chau[3], Muhammad Raziq Rahimi Kooh[3*], Roshan Thotagamuge[4*]

[1] Postgraduate Institute of Science, University of Peradeniya, Peradeniya, Sri Lanka

[2] Department of Physics, Faculty of Science, University of Peradeniya, Peradeniya, Sri Lanka

[3] Centre for Advanced Material and Energy Sciences, Universiti Brunei Darussalam, Jalan Tungku Link, Gadong BE 1410, Brunei Darussalam

[4] Department of Nano Science Technology, Faculty of Technology, Wayamba University of Sri Lanka, Kuliyapitiya 60200, Sri Lanka

*Correspondence should be addressed to:

Dr Roshan Thotagamuge, Senior Assistant Professor

Department of Nano Science Technology,

Faculty of Technology, Wayamba University of Sri Lanka,

Kuliyapitiya 60200, Sri Lanka

Email: roshan@wyb.ac.lk

Dr Muhammad Raziq Rahimi Kooh, Lecturer

Centre for Advanced Material and Energy Sciences

Universiti Brunei Darussalam

Jalan Tungku Link, Gadong BE 1410, Brunei Darussalam

Email: raziq.kooh@ubd.edu.bn

**Highlights of the work**

Novel $Cu_2$-$N_6$/Gr, $Cu_2$-$N_8$/Gr motifs investigated.

$Cu_2$-$N_6$/Gr, $Cu_2$-$N_8$/Gr are more stable than Cu-$N_4$/Gr.

$Cu_2$-$N_8$/Gr, Cu-$N_4$/Gr showed spontaneous ORR.

Inferior probability of $H_2O_2$ formation.

Maximum cell potential reached 0.49V.

**Graphical abstract of the work**

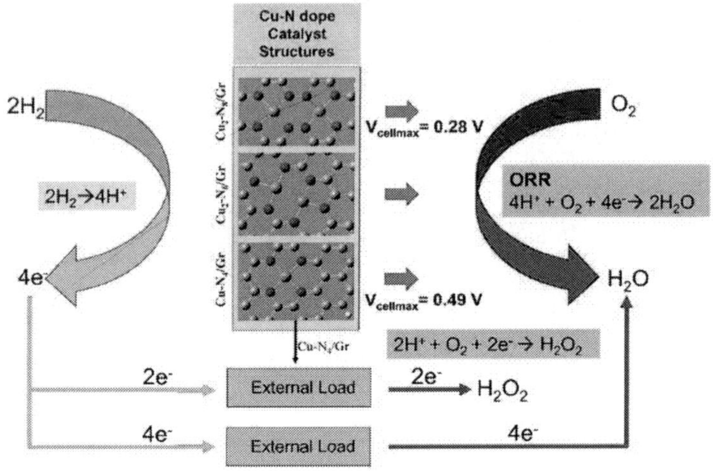

# ABSTRACT (or Blurb)

Explore the research conducted on novel Cu-nitrogen doped graphene ($Cu-N_2/Gr$) nanocomposite catalysts, specifically designed for investigating their potential as fuel cell cathode catalysts. Through rigorous density functional theory calculations, utilizing the Gaussian 09w software, this study delves into the oxygen reduction reaction (ORR) occurring on $Cu-N_2/Gr$ nanocomposite cathode catalysts in low-temperature fuel cells. Three distinct nanocomposite structures, namely $Cu_2-N_6/Gr$, $Cu_2-N_8/Gr$, and $Cu-N_4/Gr$, were meticulously examined under standard conditions (298.15 K, 1 atm) in an acidic environment, unveiling vital insights into their fuel cell properties. The comprehensive results indicate the structural stability of all three nanocomposites within the potential range of 0 - 5.87 V. By analyzing formation energy, Mulliken charge, and HOMO-LUMO energy, it becomes evident that $Cu_2-N_6/Gr$ and $Cu2-N_8/Gr$ exhibit superior structural stability. Additionally, free energy calculations reveal that only $Cu_2-N_8/Gr$ and $Cu-N_4/Gr$ structures support spontaneous ORR. Notably, the maximum cell potential under standard conditions was observed to be 0.28 V for $Cu_2-N_8/Gr$ and 0.49 V for $Cu-N_4/Gr$. This study also sheds light on the generation of $H_2O_2$, highlighting that $Cu_2-N_6/Gr$ and $Cu_2-N_8/Gr$ structures demonstrate less favorable outcomes. However, $Cu-N_4/Gr$ exhibits promising potential for $H_2O_2$ generation. In conclusion, this remarkable investigation establishes that $Cu_2-N_8/Gr$ and $Cu-N_4/Gr$ nanocomposites display a higher affinity for ORR compared to $Cu_2-N_6/Gr$. Immerse yourself in the cutting-edge world of $Cu-N_2/Gr$ nanocomposites as this study paves the way for future advancements in fuel cell technology.

# TABLE OF CONTENT

# LIST OF ABBREVIATIONS

| | |
|---|---|
| A | Electron affinity (≅-ELUMO) |
| BEs | Binding energies |
| DFT | Density functional theory |
| e | Charge transferred |
| $E_g$ | Energy gap |
| Gr | Graphene |
| $H^+$ | Hydrogen ion |
| Ha | Hartree atomic unit |
| HOMO | Highest occupied molecular orbital |
| I | Ionization potential (≅-EHOMO) |
| $K_B$ | Boltzmann's constant |
| LUMO | Lowest unoccupied molecular orbital |
| n | Number of electrons |
| $O_2$ | Oxygen |
| ORR | Oxygen reduction reaction |
| PBC | Periodic boundary condition |
| Pd | Palladium |
| PEMFCs | Proton exchange membrane fuel cells |
| PGM | Platinum group metal |
| Pt | Platinum |
| $Q_x$ | Mulliken charge of the adsorbate |
| Ry | Rydberg |
| S | Entropy |
| SE | Solvent energies |
| T | Absolute temperature |
| U | Electrode potential |
| VASP | Vienna ab-initio Simulation Package |
| $\Delta E$ | Formation energy |
| $\Delta G$ | Free energy |
| $\Delta H$ | Enthalpy |
| $\Delta Q_x$ | Mulliken charge population |
| $\Delta ZPE$ | Correction of zero-point energy |
| $\eta$ | Chemical hardness |
| $\mu$ | Chemical potential |
| $\omega$ | Electrophilicity index |

# LIST OF APPENDICES

# CHAPTER 1

## INTRODUCTION

Energy security and sustainability are the impetus for modern economics. Transformation of the energy ecosystem requires the search for cost-effective, secure, and sustainable energy technologies to replace incumbent energy sources. Fuel cells contribute well to this transformation as they evidently have their merits, particularly for compactness and zero emissions, such as proton exchange membrane fuel cells (PEMFCs) [1].

Low-temperature fuel cells operate at temperatures below 200 °C and a carbon support catalyst is deployed to improve the oxygen reduction reaction (ORR). Carbon support catalysts are not conducive to high temperature operations, leading to an increased rate of carbon corrosion and, therefore, degrading the carbon support [2]. Corrosion of the carbon support will directly affect the catalyst layer, causing the catalyst materials to agglomerate.

Generally, low-temperature commercial fuel cells use noble metals, such as Pt and Pd, as a catalyst. These noble metal catalysts can efficiently adsorb and split hydrogen into hydrogen ions [3, 4] at the anode and provide a better oxygen reduction reaction (ORR) at the cathode [5]. The damage to the catalyst via carbon corrosion, where the carbon support starts to oxidize, the catalyst particles coated on top of the carbon support then lose their support and start to agglomerate, hence, resulting in diminished fuel cell performance over a short operation period.

To reduce the cost due to carbon corrosion, cheaper alternatives are used to replace the catalyst layer. One such example is the non-Platinum group metal (non-PGM) catalysts. Non-PGM catalysts are usually produced using nitrogen atoms doped onto a graphite layer [6-8] or a layer of transition metals [9]. These are easy to produce and are much less expensive than PGM or PGM alloy catalysts such as Pt/Ni/Ir and Pt/Ni [10-12]. In addition, non-PGM have better electron conductivity. Fe/N/Gr, Co/N/Gr, and Mn/N/Gr are the most commonly used dopant [13, 14] and are studied. The Mn/N/Gr type catalyst has a more stable potential cycle than Fe; however, the catalyst is not stable after the cathode potential exceeds 0.53 V [15]. Furthermore, the Mn $N_4$/Gr catalyst has OOH dissociation at 0.38 eV [15].

For Co-based non-PGM catalysts, Co-$N_4$ has electrode stability between 0 and 1.23 V, and Co-$N_2$ is only stable below 0.45 V [16]. Furthermore, its OOH binding energy is -1.02 eV for Co-$N_4$ and -1.75 eV for Co-$N_2$ without solvent, which are more favorable than Mn. Even with the best stability of the Co-$N_4$ catalyst, research showed that Co-$N_2$ has a stronger trend towards the ORR two-electron pathway and produces $H_2O_2$ as an intermediate [16]. $H_2O_2$ causes degeneration of the electrolyte membrane.

Compared to Mn and Co, Fe-based non-PGM catalysts produce less $H_2O_2$ and this $H_2O_2$ generation can be reduced by increasing the Fe content to 1.0 wt %. The generation of $H_2O_2$ in the ORR process at 0.1 wt % for common transition metals is as follows [17], N/Gr (39.4%) > N/Gr/Ni (39.0%) > N/Gr/Mn (23.9%) > N/Gr/Co (16.8%) > N/Gr/Fe (9.6%).

The formation of $Fe(OH)_2$ during the ORR process [18] can remove Fe atoms from the carbon layer, which reduces the ORR efficiency, and the N/Gr/Fe catalyst can degenerate by oxidation [19].

Research on copper is mainly based on a single metal with nitrogen atoms as dopants [20, 21], which gives good ORR performance, and spontaneous ORR makes them more favorable for catalysts. A single metal Cu-$N_2$/Gr structure shows that it has a higher probability of following two-electron pathways to generate $H_2O_2$. In addition, there was no indication of the cathode potential of the Cu-nitrogen doped catalyst. A single structure of Cu-$N_2$/Gr cannot describe the $H_2O_2$ generation capability or the cell potential of the Cu-nitrogen doped catalyst. Therefore, other structures, such as dual-metal structures, should be considered. Research shows that Fe-based non-PGM catalysts $Fe_2$-$N_6$/Gr and $Fe_2$-$N_8$/Gr have good formation energies and ORR performance [22, 23]. Since Cu has shown better ORR performance steps than Fe [24], the $Cu_2$-$N_6$/Gr and $Cu_2$-$N_8$/Gr structures should have considerably better ORR performance than the single metal Cu-nitrogen doped catalyst or Fe-based non-PGM catalysts. Nørskov et. al. reported that Cu as a dopant with nitrogen in graphite gives ORR performance almost as Pt (111) [24] catalyst surface for the binding energy of O and OH. Furthermore, Cu shows a good binding energy of OH and OOH, indicating stable catalysts for ORR [25, 26].

The DFT method is an effective technique to study ORR performance and catalyst properties [27]. Therefore, in this study, DFT calculations were performed to determine the ORR mechanism and stability of $Cu_2$-$N_6$/Gr, $Cu_2$-$N_8$/Gr, and Cu-$N_4$/Gr electrocatalysts in an acidic medium with different potentials at 298.15 K. From these calculations, the structural stability, the binding ability of ORR steps, the break-free ability of $H_2O$, evidence of $H_2O_2$ generation, and the maximum cell potential were predicted.

# CHAPTER 2

## METHODOLOGY

The DFT calculations were performed with Gaussian 09w software using the B3LYP/3-21G basis set with non-periodic and non-dispersion interaction as Bhatt et al.[21]. Defect structure form and visualize with GaussView 6.0. as shown in **Figure 2.1**. All atoms in every structure were relaxed using optimization. A common single pristine graphene layer was developed, and it is used to develop other structures ($Cu_2$-$N_6$/Gr, $Cu_2$-$N_8$/Gr, and Cu-$N_4$/Gr) to investigate ORR considering the defect structure at the surface of the catalyst. All structures were developed to contain only pyridinic nitrogen. After optimization, zero imaginary frequency for dual atom structures and one (1) imaginary frequency for the Cu-$N_4$/Gr structure were observed. Additionally, the optimized charges of all three catalysts were zero and singlet spin for all dual atom structures. The molecules used in the ORR steps were individually optimized. Bond lengths and bond angles were measured and compared with those of previous research.

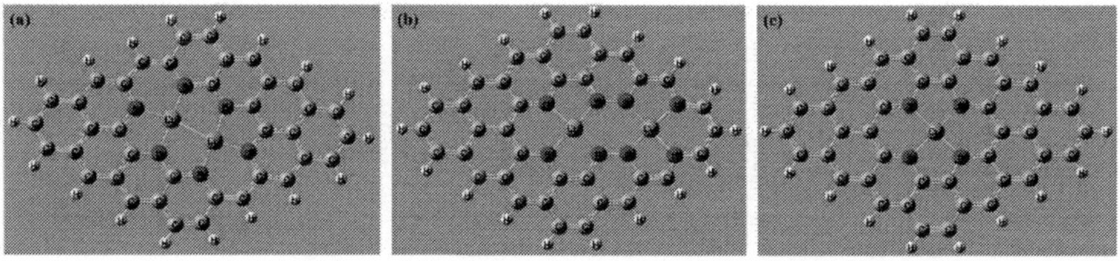


**Figure 2.1:** Top view of optimized defect structures. (a) $Cu_2$-$N_6$/Gr (b) $Cu_2$-$N_8$/Gr (c) Cu-$N_4$/Gr

Catalyst structures based on Cu confirmed their stability with the cathode potential. Stability was inspected using the formation energy (ΔE) defined accordingly eq. (1)[16, 21],

$$\Delta E = E_{graphene+(M_a-N_b)} + y\mu_C - (E_{graphene} + x\mu_N + E_M) \quad (1)$$

Here, $E_{graphene+(M_a-N_b)}$ is energy of optimized graphene layer with Cu-N defect. The $'a'$ and $'b'$ are positive integers defining selected Cu-N defects ($a$=1,2 and $b$=4,6,8). M and N are Cu and nitrogen, respectively. The $\mu_C$ and $\mu_N$ are chemical potential of carbon defined as total energy per carbon atom for defect-free graphene, and the chemical potential of nitrogen defined as half of the total energy of $N_2$ molecule, respectively. x and y are the number of nitrogen atoms added and the carbon atoms removed during the defect formation, respectively. $E_{graphene}$ is the energy of optimized pristine graphene layer. $E_M$ is the total energy of $M^{n+}$ defined as eq. (2),

$$E(M^{n+}) = E(M) - neU \quad (2)$$

where, $E(M)$ is the total energy of isolated M (M= Cu) in the gas phase and, n, e, U are the number of electron transfer (+2), electron charge, and external potential in order.

Binding energies (BEs) were calculated from the ORR steps in an acidic medium, as defined by eq. (3) [16, 20, 21],

$$BE = E_{defect+molecule} - (E_{defect} + E_{molecule}) \tag{3}$$

Here, $E_{defect+molecule}$ is the total energy of molecules adsorbed by defect graphene. $E_{defect}$ is the total energy of defect graphene configuration, and $E_{molecule}$ is isolated molecule species ($O_2$, O, $H_2O$, OOH, OH, $H_2O_2$). Negative signed binding energies ($E_{defect+molecule} < (E_{defect} + E_{molecule})$) indicate it is more favorable for molecules to be attached to the defect configuration. The formation of $H_2O_2$ during ORR is also considered.

The ORR steps in $H^+$ medium are defined as follows [16], * indicates the defect's configuration.

$* + O_2 + 4H^+ + 4e^- \rightarrow *O_2 + 4H^+ + 4e^-$

$*O_2 + 4H^+ + 4e^- \rightarrow *OOH + 3H^+ + 3e^-$

$*OOH + 3H^+ + 3e^- \rightarrow *O + H_2O + 2H^+ + 2e^-$

$*O + H_2O + 2H^+ + 2e^- \rightarrow *OH + H_2O + H^+ + e^-$

$*OH + H_2O + H^+ + e^- \rightarrow 2H_2O$

The free energies were calculated for each ORR step of all defects configurations as defined by eq. (4) [16],

$$\Delta G = \Delta E + \Delta ZPE - T\Delta S + \Delta G_U + \Delta G_{pH} + \Delta G_{field} \tag{4}$$

$\Delta E$ is the energy from the DFT calculation to the relevant reaction step, and $\Delta ZPE$ is the correction of zero-point energy is obtained from the NIST database and DFT calculations. $T$ and $S$ are the absolute temperature and entropy, respectively. Here, $T$ = 298.15 K, and the entropy value was obtained from the NIST database. $\Delta G_U = -eU$ where $U$ and $e$ are the electrode potential and the charge transferred, respectively. $\Delta G_{pH} = K_B T \times ln10 \times pH$, where $K_B$ is Boltzmann's constant and T = 298.15 K. $\Delta G_{field}$ is a contribution of the interaction of adsorbate with the local electric field in the electric double layer formed in the vicinity of the cathode, which is negligible according to a J. K. Nørskov et al. [24]. Corrections of zero-point energy and entropy values were obtained from frequency calculations for the molecule adsorb by defect graphene. Free energy *vs.* reaction coordinate graphs were drawn with different potentials (U) for each structure. Additionally, the free energy of $O_2$, 4.92 eV, was obtained from J. K. Nørskov, Yu L., et al. under standard conditions with reaction $O_2 + 2H_2 \rightarrow 2H_2O$ [24, 28].

The Mulliken charge population was applied to molecules as defined by eq. (5) [29],

$$\Delta Q_x = Q_{after} - Q_{before} \quad (5)$$

The $Q_{after}$ and $Q_{before}$ are charges of the molecule (X) after and before adsorption, respectively.

The energy gap ($E_g$), chemical hardness ($\eta$), chemical potential ($\mu$), and electrophilicity index ($\omega$) were calculated using the Koopman's principle for optimized structures as defined by eq. (6)-(9) [30-33],

$$E_g = E_{LUMO} - E_{HOMO} \quad (6)$$

$$\eta = I - A \quad (7)$$

$$\mu = -\frac{(I+A)}{2} \quad (8)$$

$$\omega = \frac{\mu^2}{2\eta} \quad (9)$$

Here, I and A are the ionization potential ($\cong$-$E_{HOMO}$) and the electron affinity ($\cong$-$E_{LUMO}$), respectively.

# CHAPTER 3

## RESULTS AND DISCUSSION

### 3.1 Formation Energy

The formation energies of $Cu_2$-$N_6$/Gr, $Cu_2$-$N_8$/Gr, and Cu-$N_4$/Gr structures were -23.60, -23.49 and -13.22 eV in order at zero potential (U = 0) as shown in **Table 3.1**. These values were only applied under open circuit conditions, and they changed with different cathode potentials according to eq. (1)[16, 21]. **Table 3.1** shows that the formation energy increased with increasing external potentials. A negative sign means energy value gain when a structure forms. Therefore, the negative formation energies are favorable for the formation of stable structures, and the formation energies of all the structures studied remain negative between the potential range of 0 to 5.87 V. The critical cell potential of negative formation energies was observed at 5.89, 5.87, and 6.60 V for $Cu_2$-$N_6$/Gr, $Cu_2$-$N_8$/Gr, and Cu-$N_4$/Gr, respectively, see **Table 3.1**. The critical U value at zero formation energy demarcates the stability/instability of the cathode. The structures forming stability follow the increasing order of Cu-$N_4$/Gr < $Cu_2$-$N_8$/Gr < $Cu_2$-$N_6$/Gr, as indicated by the external potential, U, at zero.

**Table 3.1:** Formation energies for $Cu_2$-$N_6$/Gr, $Cu_2$-$N_8$/Gr, and Cu-$N_4$/Gr structures with different external potentials (U).

| DEFECT | EXTERNAL POTENTIAL (U) / V | FORMATION ENERGY / eV |
|---|---|---|
| **$Cu_2$-$N_6$/Gr** | 0.00 | -23.60 |
| | 1.00 | -19.60 |
| | 2.00 | -15.60 |
| | 5.89 | -0.04 |
| | 5.90 | 0.00 |
| **$Cu_2$-$N_8$/Gr** | 0.00 | -23.49 |
| | 1.00 | -19.49 |
| | 3.00 | -11.49 |
| | 5.87 | -0.01 |
| | 5.88 | 0.03 |
| **Cu-$N_4$/Gr** | 0.00 | -13.22 |
| | 1.00 | -11.22 |
| | 4.00 | -5.22 |
| | 6.60 | -0.02 |
| | 6.61 | 0.00 |

## 3.2 Binding Ability

The binding energies of ORR intermediates were calculated in an acidic medium involving $O_2$, O, $H_2O$, OOH, OH and $H_2O_2$ molecules (**Table 3.2**), and with and without the water solvent energies obtained from Sha et. al.'s work [34]. Kattel et. al. and Bhatt et. al. reported that binding energies of $O_2$ to the structure of Co-$N_4$/Gr and Cu-$N_2$/Gr were -0.67 and -2.90 eV, respectively. As a first ORR step, $O_2$ binding to the defect shown in this study is significantly high, except for the Cu-$N_4$/Gr structure in relation to Cu-$N_2$/Gr [16, 21]. Structure-wise Cu-$N_2$/Gr has a higher $O_2$ binding energy than Cu-$N_4$/Gr, also, $Cu_2$-$N_6$/Gr has a higher binding energy compared to the $Cu_2$-$N_8$/Gr structure. This indicates that adding more nitrogen atoms to Cu did not create a good $O_2$ adsorbent defect. The efficiency of the $O_2$ adsorbent in decreasing order is as follows, $Cu_2$-$N_6$/Gr > $Cu_2$-$N_8$/Gr > Cu-$N_4$/Gr.

The binding energies of OH and OOH on the $Cu_2$-$N_6$/Gr structure are shown to be -5.12 and -4.13 eV and -3.81 and -3.27 eV, respectively, for binding onto the $Cu_2$-$N_8$/Gr structure. The binding energies for OH and OOH to the Cu-$N_2$/Gr structure were reported to be -2.68 and -1.81 eV, respectively, while for the Cu-$N_4$/Gr structure were -1.99 and -1.70 eV [21]. The OH and OOH binding energies also signify the above N atoms and the idea of a stable defect. The energy values, shown in **Table 3.2**, also suggest that the $Cu_2$-$N_6$/Gr defect supports high ORR molecule binding.

**Table 3.2:** Binding energies (BE) of $Cu_2$-$N_6$/Gr, $Cu_2$-$N_8$/Gr, Cu-$N_4$/Gr structures with and without solvent energies (SE) [34]. (01) and (02) in the $Cu_2$-$N_6$/Gr defect represent the different ORR paths 03-04-06-08 and 01-05-08 in **Figure 3.1**, respectively.

| DEFECT | MOLECULE | BE (WITHOUT SOLVENT) / eV | SE / eV | BE (WITH SOLVENT) /eV |
|---|---|---|---|---|
| **$Cu_2$-$N_6$/Gr (01)** | O+O | -3.54 | -0.41 | -3.95 |
| | OOH | -4.13 | -0.47 | -4.60 |
| | O | -7.53 | -0.70 | -8.23 |
| | OH | -5.12 | -0.38 | -5.50 |
| | $H_2O$ | -1.30 | - | - |
| | OH+OH (H2O2) | -5.44 | - | - |
| **$Cu_2$-$N_6$/Gr (02)** | O+O | -3.78 | -0.41 | -4.19 |
| | O+OH | -9.52 | -1.08 | -10.60 |
| | OH+OH | -6.52 | -0.76 | -7.28 |
| | OH | -5.13 | -0.38 | -5.51 |
| | $H_2O$ | -1.30 | - | - |
| **$Cu_2$-$N_8$/Gr** | $O_2$ | -2.79 | -0.41 | -3.20 |
| | OOH | -3.27 | -0.47 | -3.74 |
| | OH+OH | -5.34 | -0.76 | -6.10 |
| | OH | -3.81 | -0.38 | -4.19 |
| | $H_2O$ | -0.98 | - | - |
| **Cu-$N_4$/Gr** | $O_2$ | -2.76 | -0.41 | -3.17 |
| | OOH | -1.70 | -0.47 | -2.17 |
| | O | -5.15 | -0.70 | -5.85 |
| | OH | -1.99 | -0.38 | - |
| | $H_2O$ | -1.23 | - | - |
| | $H_2O_2$ | -0.92 | - | - |

There are two possibilities of ORR pathways for $Cu_2$-$N_6$/Gr as shown in**Figure 3.1** and **Table 3.2**. Both pathways show that the $O_2$ adsorbent is high relative to the other structures being considered in this study. For the structure of $Cu_2$-$N_6$/Gr, the 1-5-8 pathway has a higher $O_2$ adsorbent affinity than the 3-4-6-8 pathway. For both the 1-5-8 and 3-4-6-8 pathways, the initial step is the same but having different binding energies may cause the ORR to take two different paths. **Figure 3.2 (a - e)** show all five ORR steps of 1-5-8 pathway with intermediates of *$O_2$, *O+OH, *OH+OH, *OH, *$H_2O$ and **Figure 3.3 (a - e)** show all five ORR steps of 3-4-6-8 pathway with intermediates of *$O_2$,*OOH, *O, *OH, *$H_2O$ respectively. **Figure 3.3 (f)** shows the result of trying to form $H_2O_2$ from *OOH intermediate.

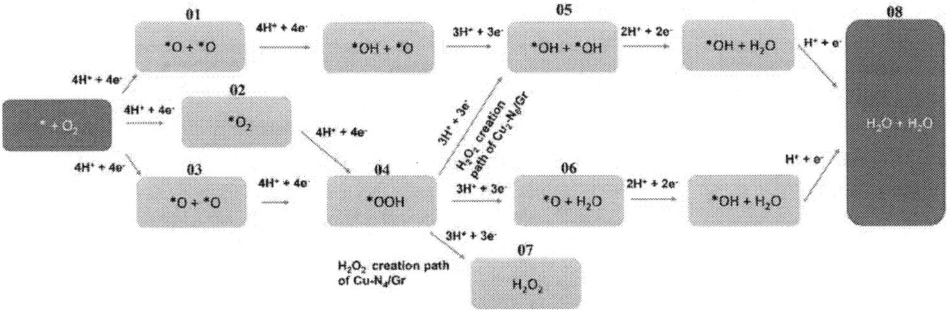


**Figure 3.1:** ORR paths for $Cu_2$-$N_6$/Gr, $Cu_2$-$N_8$/Gr, Cu-$N_4$/Gr structures

The difference in free energy ($\Delta G$) for $O_2$ binding was -3.59 and -3.58 eV for $Cu_2$-$N_6$/Gr (01) and $Cu_2$-$N_6$/Gr (02), respectively. These $\Delta G$ values indicate that the $Cu_2$-$N_6$/Gr structure favors both pathways, as shown by the $O_2$ adsorption step.

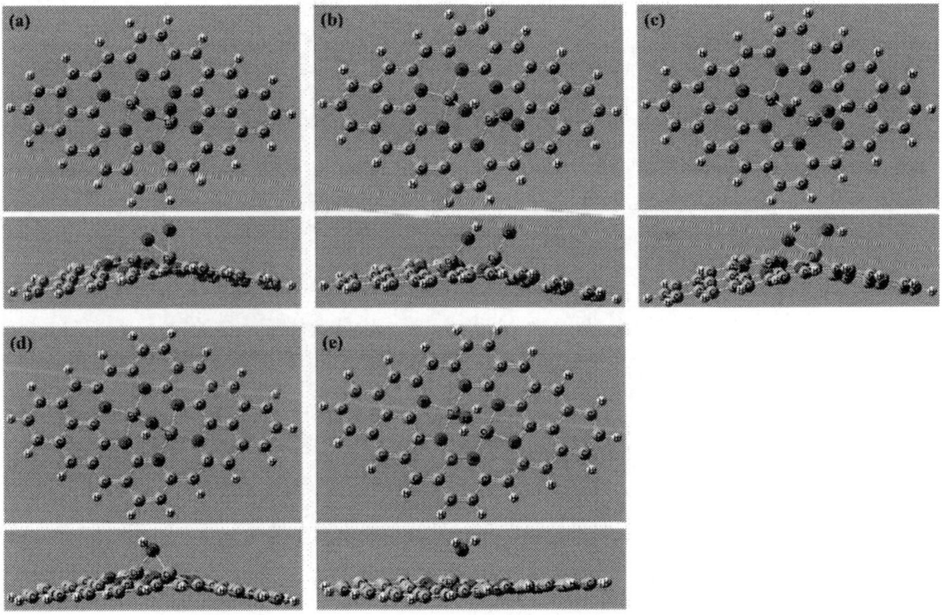


**Figure 3.2:** Top and side view of optimized ORR steps for $Cu_2$-$N_6$/Gr (02) structure. (a) $O_2$ (b) O+OH (c) OH+OH (d) OH (e) $H_2O$ molecules binding on $Cu_2$-$N_6$/Gr (02) defect. (White, grey, blue, orange and red spheres represent hydrogen (H), carbon (C), nitrogen (N), copper (Cu) and oxygen atoms (O), respectively)

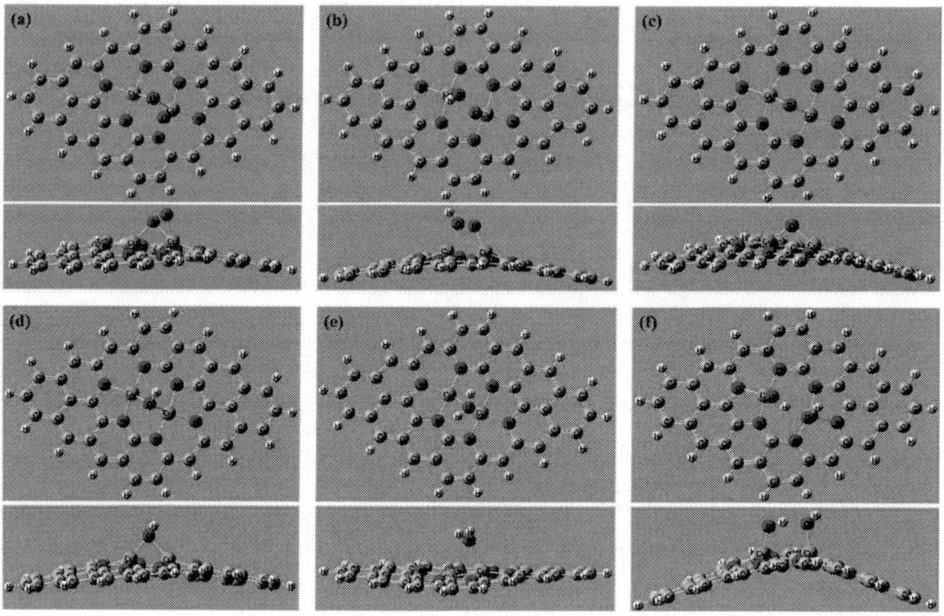


**Figure 3.3:** Top and side view of optimized ORR steps for $Cu_2$-$N_6$/Gr (01) structure. **(a)** $O_2$ **(b)** OOH **(c)** O **(d)** OH **(e)** $H_2O$ **(f)** dual OH molecules form and binding instead of $H_2O_2$ on $Cu_2$-$N_6$/Gr (01) defect. (White, grey, blue, orange and red spheres represent hydrogen (H), carbon (C), nitrogen (N), copper (Cu) and oxygen atoms (O), respectively)

The Cu-O bond length of $O_2$ was 1.82, 1.89, and 1.98 Å for $Cu_2$-$N_6$/Gr, $Cu_2$-$N_8$/Gr, and Cu-$N_4$/Gr, respectively as shown in **Table 3.3**. The results show that the most important first step of ORR favors the $Cu_2$-$N_6$/Gr structure. Xiao et. al. and Isaacs et. al. reported Cu-O bond lengths ranging from 2.04 to 2.59 Å [20, 35]. However, in this study, the distance of Cu-O bond lengths is less than 2 Å for both the $Cu_2$-$N_6$/Gr and $Cu_2$-$N_8$/Gr structures, and this may be due to the binding of the O atom to dual Cu atoms as shown in **Figures 3.2 - 3.4**. Here,

**Figure 3.4 (a - e)** shows optimized ORR steps for $Cu_2$-$N_8$/Gr structure with *$O_2$, *OOH, *OH+OH, *OH, *$H_2O$ intermediates respectively. For the Cu-$N_4$/Gr structure, the Cu-O bond length is comparable to the minimum value of the Cu-O bond range studied previously. Furthermore, O-O bond length of $O_2$ binding to Cu-$N_4$/Gr structure shows similarity to Cu-$N_2$/Gr structure [21]. Additionally, a prior investigation using VASP (Vienna ab-initio Simulation Package) [20] on Cu-$N_4$/Gr indicated that the Cu-O bond lengths for the $O_2$ and O intermediates were 2.22 Å and 1.67 Å, respectively, with $O_2$ exhibiting weaker binding compared to O, potentially attributed to varying basis set configurations employed in the study. O-O bond length of $O_2$ binding in $Cu_2$-$N_6$/Gr and $Cu_2$-

$N_8$/Gr structures shows a slight increase in bond length. This increase in the bond length may be due to π bonding electron on the adsorption side of O weakened by the Cu-O bond.

**Table 3.3:** Shortest distance (d) between Cu-O, O-O atoms of $Cu_2$-$N_6$/Gr, $Cu_2$-$N_8$/Gr, Cu-$N_4$/Gr structures in angstroms (Å). (01) and (02) in the $Cu_2$-$N_6$/Gr defect represent the different ORR paths 03-04-06-08 and 01-05-08 in ***Figure 3.1***, respectively.

| DEFECT | MOLECULE | $d_{O\text{-}O}$ /Å | $d_{Cu\text{-}O}$ /Å |
|---|---|---|---|
| **$Cu_2$-$N_6$/Gr (01)** | O+O | 1.50 | 1.82 |
| | OOH | 1.55 | 1.84 |
| | O | - | 1.74 |
| | OH | - | 1.89 |
| | $H_2O$ | - | 2.09 |
| | OH+OH ($H_2O_2$) | 2.48 | 1.75,1.85 |
| **$Cu_2$-$N_6$/Gr (02)** | O+O | 1.49 | 1.82 |
| | O+OH | 2.52 | 1.83 |
| | OH+OH | 2.45 | 1.79 |
| | OH | - | 1.89 |
| | $H_2O$ | - | 2.09 |
| **$Cu_2$-$N_8$/Gr** | $O_2$ | 1.46 | 1.89 |
| | OOH | 1.55 | 1.93 |
| | OH+OH | 2.66 | 1.81,1.86 |
| | OH | - | 2.05 |
| | $H_2O$ | - | 2.20 |
| **Cu-$N_4$/Gr** | $O_2$ | 1.38 | 1.98 |
| | OOH | 1.52 | 1.87 |
| | O | - | 1.88 |
| | OH | - | 1.84 |
| | $H_2O$ | - | 2.08 |
| | $H_2O_2$ | 1.54 | 2.18 |

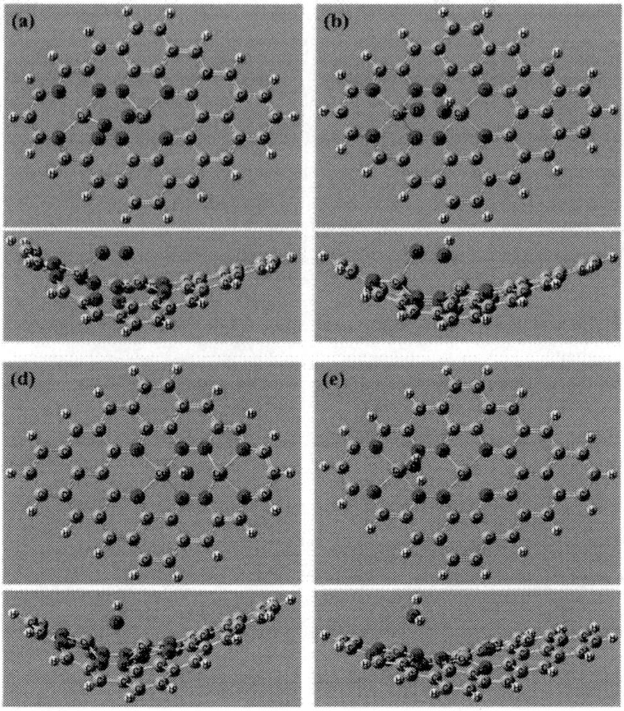


**Figure 3.4:** Top and side view of optimized ORR steps for $Cu_2$-$N_8$/Gr structure. **(a)** $O_2$ **(b)** OOH **(c)** OH+OH **(d)** OH **(e)** $H_2O$ molecules binding on $Cu_2$-$N_8$/Gr defect. (White, grey, blue, orange and red spheres represent hydrogen (H), carbon (C), nitrogen (N), copper (Cu) and oxygen atoms (O), respectively)

When molecule binding occurs, the Cu atom protrudes from the graphene layer. According to the optimized structures, most of the protruding Cu atom occur due to the binding of the O atom or the OH molecule (**Figures 3.3, 3.5, 3.6**), and this is consistent with works reported by Xiao et. al. and Bhatt et. al. [20, 21]. The final ORR step involves $H_2O$ binding. The $H_2O$ binding strength follows a decreasing order with respect to the defect structures: $Cu_2$-$N_6$/Gr> Cu-$N_4$/Gr > $Cu_2$-$N_8$/Gr. Based on the results of this study, it has been confirmed that the attachment of $H_2O$ to the Cu atom on the $Cu_2$-$N_8$/Gr defect structure is weakest, as evidenced by the 2.20 Å Cu-O bond length, when compared to the other defect structures investigated. On the other hand, $Cu_2$-$N_6$/Gr is shown to have a high $H_2O$ adsorbent possibility but also to have a better adsorption ability for $O_2$ than $H_2O$. The binding energies indicate that Cu-$N_4$/Gr has less ability to adsorb $H_2O$ when compared to $Cu_2$-$N_6$/Gr. The bond length to $H_2O$ is similar for both defects. The binding energies of *OH, *OOH and *O have significantly higher values, indicating that all three defects could be considered as possible stable active sites for ORR.

### 3.3 Effects of the Mulliken Charge

The Mulliken charge for the adsorbate of the optimized structures is shown in **Table 3.4**. The Mulliken charge population ($\Delta Q_x$) was applied to both O atoms of the $O_2$ intermediate as eq. (5) for all structures. $\Delta Q_x$ values for $Cu_2$-$N_6$/Gr (01) -0.063 e and 0.239 e, $Cu_2$-$N_6$/Gr (02) -0.141 e and -0.077 e, $Cu_2$-$N_8$/Gr 0.02 e and -0.151 e, Cu-$N_4$/Gr -0.024 e. Here, the negative sign indicates an electron loss of the intermediate [29]. Therefore, the electron loss of the O atom could be considered as the electron gain of the Cu atom, which could explain the increased binding strength. The decreasing order of the charge population bonding strength is $Cu_2$-$N_6$/Gr > $Cu_2$-$N_8$/Gr > Cu-$N_4$/Gr, suggesting that that $Cu_2$-$N_6$/Gr has the highest $O_2$ adsorption ability. Additionally, this observation is consistent with the decreasing order of $O_2$ adsorbent efficiency predicted using the binding energy. For $H_2O$, the Mulliken charge population values for are 0.027 e, 0.058 e, 0.008 e and -0.054 e for $Cu_2$-$N_6$/Gr (01), $Cu_2$-$N_6$/Gr (02), $Cu_2$-$N_8$/Gr and Cu-$N_4$/Gr, respectively. These values signify the binding strength of $H_2O$ in decreasing order as Cu-$N_4$/Gr > $Cu_2$-$N_8$/Gr > $Cu_2$-$N_6$/Gr (01) > $Cu_2$-$N_6$/Gr (02). Only Cu-$N_4$/Gr and $Cu_2$-$N_8$/Gr are compared with the above $H_2O$ adsorbent strength decreasing order explained by the binding energy. The difference in $H_2O$ adsorbent strength order between the Mulliken charge population and the binding energy may be due to the O atom drawing electrons from H atoms of $H_2O$, altering the Mulliken charge values of O atom. Based on the Mulliken charges of $Cu_2$-$N_6$/Gr, it has the weakest bond strength on $H_2O$, allowing $H_2O$ to break free from the catalyst easily. The Mulliken charges of optimized structures in **Table 3.4** indicate that the bonding capability is dependent on the difference of $Q_x$ and $Q_{Cu}$ and bond length $d_{Cu\text{-}O}$. Strong bonding is favored by a large difference in $Q_x$ and $Q_{Cu}$ and a short bond length $d_{Cu\text{-}O}$.

**Table 3.4:** The Mulliken charge of the adsorbate ($Q_x$) and the Mulliken charge of the Cu site ($Q_{Cu}$) of $Cu_2$-$N_6$/Gr, $Cu_2$-$N_8$/Gr, Cu-$N_4$/Gr structures for the shortest distance of the optimized structure. (01) and (02) in the $Cu_2$-$N_6$/Gr defect represent the different ORR paths 03-04-06-08 and 01-05-08 in **Figure 3.1**, respectively. Two Mulliken charge values represent the atom order of the molecule column. Here, e = 1.602 x $10^{-19}$ C.

| DEFECT | MOLECULE | $Q_x$ /e | $Q_{Cu}$ /e |
|---|---|---|---|
| **$Cu_2$-$N_6$/Gr (01)** | O+O | -0.223, -0.365 | 0.979 |
| | OOH | -0.328 | 0.934 |
| | O | -0.592 | 1.087 |
| | OH | -0.643 | 0.952 |
| | $H_2O$ | -0.573 | 0.937 |
| | OH+OH ($H_2O_2$) | -0.618, -0.602 | 0.947, 0.744 |
| **$Cu_2$-$N_6$/Gr (02)** | O+O | -0.367, -0.220 | 1.010 |
| | O+OH | -0.368, -0.661 | 0.631, 1.051 |
| | OH+OH | -0.665, -0.605 | 1.060, 0.897 |
| | OH | -0.640 | 0.949 |
| | $H_2O$ | -0.573 | 0.937 |

| | | | |
|---|---|---|---|
| **$Cu_2$-$N_8$/Gr** | $O_2$ | -0.273 | 0.967 |
| | OOH | -0.303 | 0.993 |
| | OH+OH | -0.606, -0.612 | 0.919,0.845 |
| | OH | -0.615 | 1.015 |
| | $H_2O$ | -0.599 | 1.068 |
| **Cu-$N_4$/Gr** | $O_2$ | -0.195 | 0.996 |
| | OOH | -0.312 | 0.922 |
| | O | -0.312 | 0.905 |
| | OH | -0.581 | 0.857 |
| | $H_2O$ | -0.587 | 1.011 |
| | $H_2O_2$ | -0.319 | 0.999 |

### 3.4 Possibility of Hydrogen Peroxide Formation

The $Cu_2$-$N_6$/Gr structure follows the ORR steps on both the 01-05-08 and 03-04-06-08 paths (**Figure 3.1**). The 03-04-06-08 path has the possibility of generating $H_2O_2$ due to the formation of *OOH intermediate with an additional H atom (

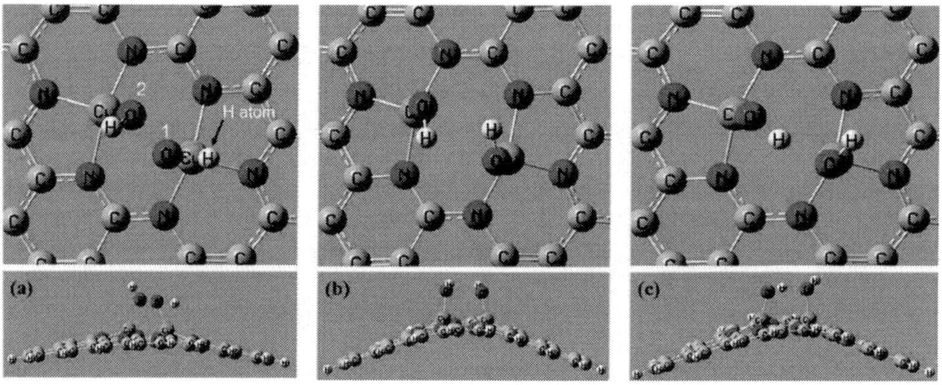


**Figure** 3.5 **(a)**), but optimization shows instead of $H_2O_2$ generation, it generates a separate *OH intermediate (

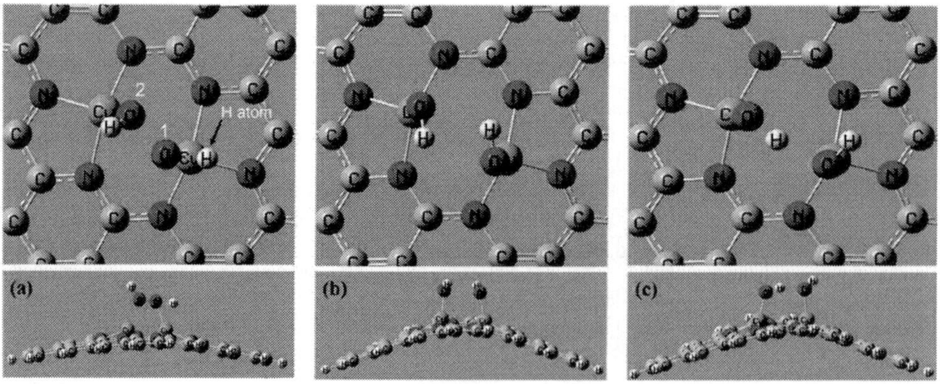

**Figure** 3.5 (**c**)) and follows the 01-05-08 path from step 05 (

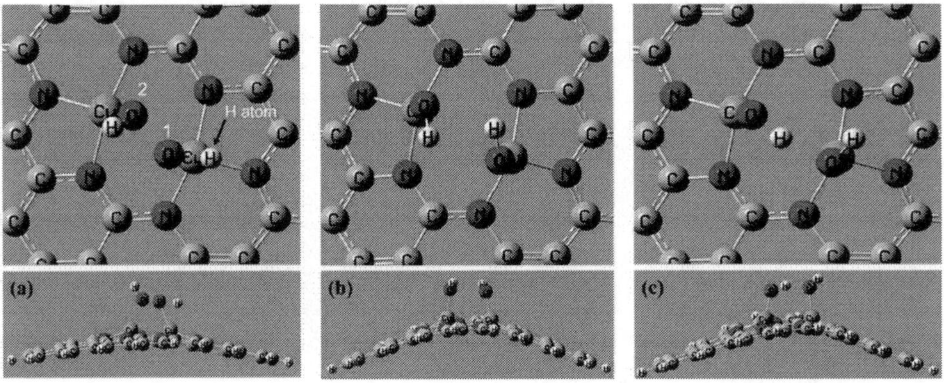


**Figure** 3.5 **(a - c)** and **Figure 3.3 (f)**).

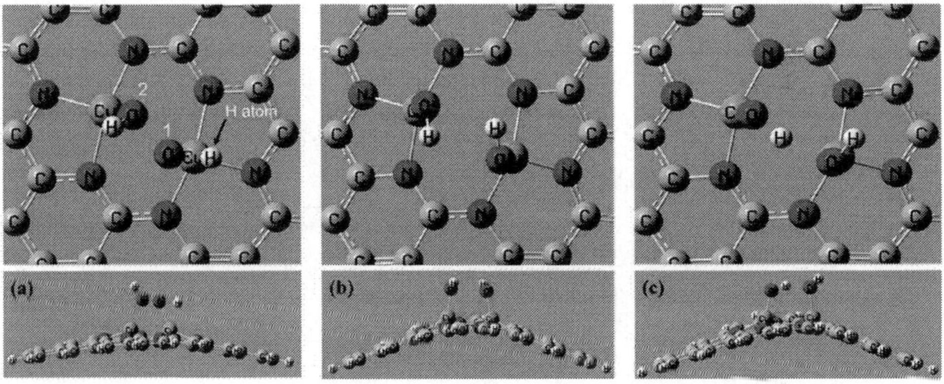


**Figure 3.5:** $Cu_2$-$N_6$/Gr (01) structure (a) H atom binds with $O_1$ atom of OOH to form $H_2O_2$ (b) Middle state of optimization (c) Final optimization. (White, grey, blue, orange and red spheres represent hydrogen (H), carbon (C), nitrogen (N), copper (Cu) and oxygen atoms (O), respectively)

Separate generation of the OH molecule may be due to the attraction of electrons by the Cu atoms. According to Mulliken charges, N atoms lose less electrons to Cu atoms of $Cu_2$-$N_6$/Gr on demand of electrons compared to Cu-$N_4$/Gr. Therefore, the Cu atom of $Cu_2$-$N_6$/Gr may try to replenish its electron demand from O atom charges by forming a strong Cu-O bond (

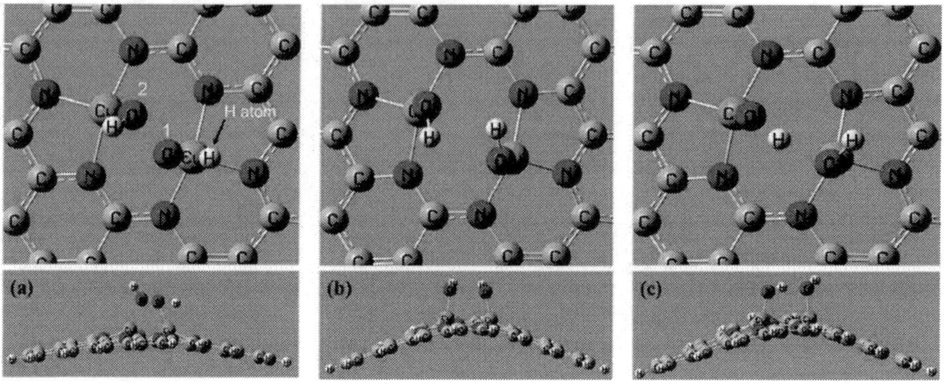


**Figure 3.5 (b - c)**) unlike Cu-$N_4$/Gr which form $H_2O_2$. Furthermore, this scenario could be explained as $\pi$ bonding electron on O atoms that attracts to each Cu atom. If O atoms largely lose their electron charge density, that could cause the O-O bond to break and generate *2OH instead of $H_2O_2$ from *OOH.

Nallathambi et. al. reported that $H_2O_2$ generation could stop the ORR four-electron path halfway [36]. The ORR steps of an acidic medium show that a two-electron pathway is needed for the generation of $H_2O_2$. However, the optimization did not produce a two-electron pathway for the $Cu_2$-$N_6$/Gr structure. $Cu_2$-$N_8$/Gr structures only follow the ORR 02-04-05-08 path (**Figure 3.1**). It is a four-electron pathway without $H_2O_2$ generation. The *OOH intermediate on its ORR path is more favorable to form dual OH over $H_2O$ (

**Figure 3.4 (c)** and

**Figure 3.6 (a - c)**). The binding of the H atom (01) to the second O atom (

**Figure 3.6 (a)**) causes the breakage of the O-O bond due to $\pi$ electron deviation. This situation creates the $H_2O$ molecule short time. However, the formation of *2OH is made possible by the attraction of electrons from the O atom of $H_2O$ by the Cu atom caused by weak Cu - N bonding. This in turn would weaken one of the O-H bonds and lead to the release of H to react to form OH (

**Figure 3.6 (b)**). Since the *OOH of $Cu_2$-$N_8$/Gr structure creates the *2OH configuration by H atoms binding to both O atoms (

**Figure 3.6 (c)**), the possibility of $H_2O_2$ formation is reduced.

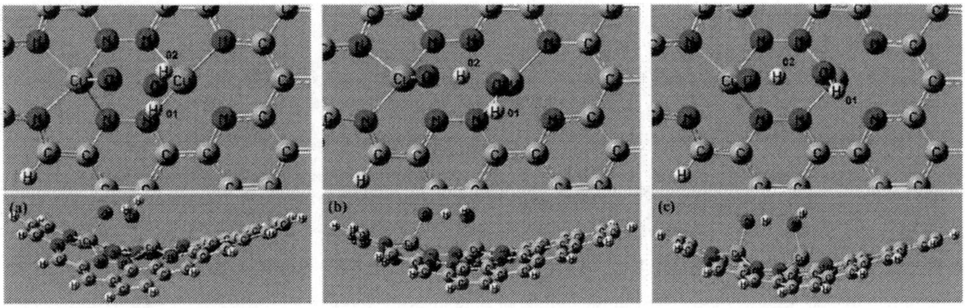

**Figure 3.6:** $Cu_2$-$N_8$/Gr structure **(a)** H atom (01) bind with second O atom of $O_1O_2H$ molecule to form $H_2O$ **(b)** H atom (01) binding to second O atom and previously bonded H atom (02) break its bond and travel to first O atom **(c)** Final optimization, forming dual OH molecules instead of $H_2O_2$. (White, grey, blue, orange and red spheres represent hydrogen (H), carbon (C), nitrogen (N), copper (Cu) and oxygen atoms (O), respectively)

The ORR associated with the Cu-$N_4$/Gr structure follows only the 02-04-06-08 path (**Figure 3.1**) with *$O_2$, *OOH, *O, *OH, *$H_2O$ intermediates respectively (**Figure 3.7 (a - e)**). The optimized structures show the possibility of $H_2O_2$ generation (**Figure 3.7 (f)**). With a binding energy of -0.92 eV and a Cu-O bond length of 2.18 Å, $H_2O_2$ can easily break free from the Cu-$N_4$/Gr defect (**Figure 3.7 (f)**). The O-O bond in the simulation is shown to be stable, as suggested by Bhatt et. al. [21]. The Mulliken charge population value of $H_2O_2$, 0.026 e, indicates that $H_2O_2$ has the ability to break free from the catalyst surface as $H_2O$. Since $H_2O_2$ directly participates in the degeneration of the fuel cell membrane [37, 38], the structure of Cu-$N_4$/Gr can significantly reduce the efficiency of the fuel cell. Additionally, the two-electron pathway of $H_2O_2$ generation can reduce the current density of the fuel cell.

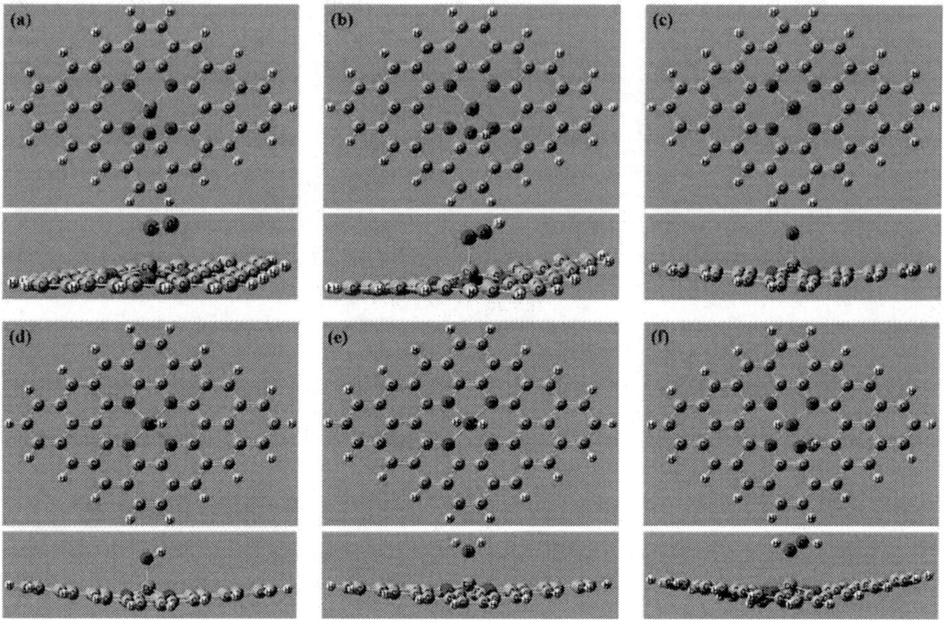


**Figure 3.7:** Top and side view of optimized ORR steps for Cu-$N_4$/Gr structure. (a) $O_2$ (b) OOH (c) O (d) OH (e) $H_2O$ (f) $H_2O_2$ molecules binding on Cu-$N_4$/Gr defect. (White, grey, blue, orange and red spheres represent hydrogen (H), carbon (C), nitrogen (N), copper (Cu) and oxygen atoms (O), respectively)

## 3.5 HOMO and LUMO Energy Calculations

HOMO and LUMO energies were calculated for every step of the ORR and structural defects to further investigate the binding energy. Energy gap ($E_g$), chemical hardness($\eta$), chemical potential ($\mu$), and electrophilicity index ($\omega$) of *$O_2$, *$H_2O$, *$H_2O_2$, intermediates and defect structures were calculated as shown in **Table 3.5** using eqs. (6)-(9) [29, 31, 32].

**Table 3.5:** Energy gap ($E_g$), chemical hardness($\eta$), chemical potential ($\mu$), and electrophilicity index ($\omega$) values of $*O_2$, $*H_2O$, $*H_2O_2$ intermediates and defect structures.

| DEFECT | INTERMEDIATE/ STRUCTURE | $E_g$ / eV | $\eta$ / eV | $\mu$ / eV | $\omega$ / eV |
|---|---|---|---|---|---|
| **$Cu_2$-$N_6$/Gr (01)** | $Cu_2$-$N_6$/Gr | 0.04 | 0.04 | -0.13 | 0.23 |
| | O+O | 0.03 | 0.03 | -0.14 | 0.33 |
| | $H_2O$ | 0.03 | 0.03 | -0.12 | 0.23 |
| | $H_2O_2$ | - | - | - | - |
| **$Cu_2$-$N_6$/Gr (02)** | O+O | 0.03 | 0.03 | -0.14 | 0.35 |
| | $H_2O$ | 0.03 | 0.03 | -0.12 | 0.22 |
| | $H_2O_2$ | - | - | - | - |
| **$Cu_2$-$N_8$/Gr** | $Cu_2$-$N_8$/Gr | 0.05 | 0.05 | -0.12 | 0.15 |
| | $O_2$ | 0.04 | 0.04 | -0.13 | 0.19 |
| | $H_2O$ | 0.05 | 0.05 | -0.11 | 0.13 |
| | $H_2O_2$ | - | - | - | - |
| **Cu-$N_4$/Gr** | Cu-$N_4$/Gr | 0.03 | 0.03 | -0.13 | 0.27 |
| | $O_2$ | 0.04 | 0.04 | -0.13 | 0.19 |
| | $H_2O$ | 0.04 | 0.04 | -0.11 | 0.16 |
| | $H_2O_2$ | 0.03 | 0.03 | -0.12 | 0.21 |

According to Eqs. 6 and 7 $\eta$ and $E_g$ are directly associated with each other. Chemical hardness elucidates the stability and reactivity of the structure. Therefore, stable structures are less reactive, and high-reactive structures are less stable [8, 29, 31]. The defects can be ranked in decreasing order of chemical hardness as $Cu_2$-$N_8$/Gr > $Cu_2$-$N_6$/Gr > Cu-$N_4$/Gr. This suggests that the reactivity of Cu-$N_4$/Gr structure is higher than other structures, but it is also less stable. This reactivity could cause electrons in the valence band to be easily excited to occupy the conduction band. The reactivity order is confirmed by the formation energy shown in **Table 3.1**. The $Cu_2$-$N_8$/Gr structure is more stable than others and is the least reactive of the three defect structures, and $Cu_2$-$N_6$/Gr has medium structural stability and reactivity according to chemical hardness values.

The chemical hardness of the oxygen bonding decreasing order $Cu_2$-$N_8$/Gr = Cu-$N_4$/Gr > $Cu_2$-$N_6$/Gr indicates that $Cu_2$-$N_6$/Gr is less stable and more reactive than the other two structures. If the ORR first step is less stable and has high reactivity, the probability of achieving the next ORR step is high. Therefore, $Cu_2$-$N_6$/Gr defect is more suitable to complete the ORR first step. From previous calculations, the binding energy of the second step for ORR of $Cu_2$-$N_6$/Gr structure is higher than that of the first step, and this collaborates well with the chemical hardness data. The chemical hardness of the $H_2O$ bond in decreasing order is as follows: $Cu_2$-$N_8$/Gr > Cu-$N_4$/Gr > $Cu_2$-$N_6$/Gr. This indicates that the $H_2O$ bonded to the $Cu_2$-$N_8$/Gr defects is more stable than the other defects. DFT calculations also show that the longest bond length of Cu-O and the highest binding energy would increase the probability of breaking $H_2O$ from the $Cu_2$-$N_8$/Gr defect sites. In the case of $H_2O$ bonding to the highly reactive $Cu_2$-$N_6$/Gr defect, this would result in oxygen evolution reaction (OER) [20, 39].

The chemical potential is directly related to Mulliken's electronegativity [31]. The chemical reactivity of the structure increases with decreasing chemical potential. The chemical potential decreases in the following order of defects: $Cu_2$-$N_8$/Gr > $Cu_2$-$N_6$/Gr = Cu-$N_4$/Gr. This shows that the chemical reactivity of the $Cu_2$-$N_6$/Gr and Cu-$N_4$/Gr structures is slightly higher than that of $Cu_2$-$N_8$/Gr. Furthermore, the chemical reactivity of the oxygen-bonded structures is the highest for $Cu_2$-$N_6$/Gr, confirming the previous chemical hardness evaluation of $Cu_2$-$N_6$/Gr.

Since electrophilicity is the electron accepting capacity from the surrounding atoms to form a stable energy state, and this relates directly to the structural stability. The electrophilicity of the studied defects has the following decreasing order, Cu-$N_4$/Gr > $Cu_2$-$N_6$/Gr > $Cu_2$-$N_8$/Gr, and this also confirms with the reactivity order of the structures in terms of chemical hardness. The oxygen-bonded $Cu_2$-$N_6$/Gr shows a higher electron acceptation than the other structures, hence, indicating that the $Cu_2$-$N_6$/Gr structure favors electron adsorption to form a stable structure. Two ORR steps of $Cu_2$-$N_6$/Gr that generate different paths could occur to this reactivity. $Cu_2$-$N_8$/Gr and Cu-$N_4$/Gr structures show similar electrophilicity values for $O_2$ bonding. The stability of the Cu-$N_4$/Gr structure is attributed to the four nitrogen atoms that donate electrons to the Cu atom. As compared to the $Cu_2$-$N_6$/Gr structure, which has six nitrogen atoms, the Cu-$N_4$/Gr structure is more stable (**Table 3.1**). The $Cu_2$-$N_8$/Gr structure has twice as many nitrogen atoms as Cu-$N_4$/Gr, but its electrophilicity value is likely to be low, similar to that of Cu-$N_4$/Gr. Additionally, $H_2O$ bonded $Cu_2$-$N_8$/Gr showed the least electrophilicity value for the last ORR step, indicating a low probability that a further reaction will take place.

### 3.6 Free Energy

Free energy values were calculated for the three defect structures with possible ORR pathways in an acidic medium as shown in **Figure 3.8 – 3.11**. $Cu_2$-$N_6$/Gr shows a spontaneous reaction (downhill) for the first two steps of ORR with cell potential (U) 0 to 1.00 V

Figure 3.8: Free energy of possible ORR path **for** Cu2N6/Gr (01) defect structure with different potentials in an acidic medium. and **Figure 3.9**). The second to the third steps of the ORR for all potentials show uphill. From the third step to the last downhill configuration was found again for 0V-1.00V. This indicates that a large energy barrier is present between the second step and third step, which prevents the spontaneous reaction of ORR to be completed. The highest reaction step barrier, 5.34 and 5.11 eV, are shown by $Cu_2$-$N_6$/Gr, for both (01) and (02) paths at U = 0 V, respectively, in *O-O hydrogenate by attaching the H atom to form *OOH and *O-OH. In summary, even with good binding energies, both ORR paths of $Cu_2$-$N_6$/Gr structure did not show spontaneity.

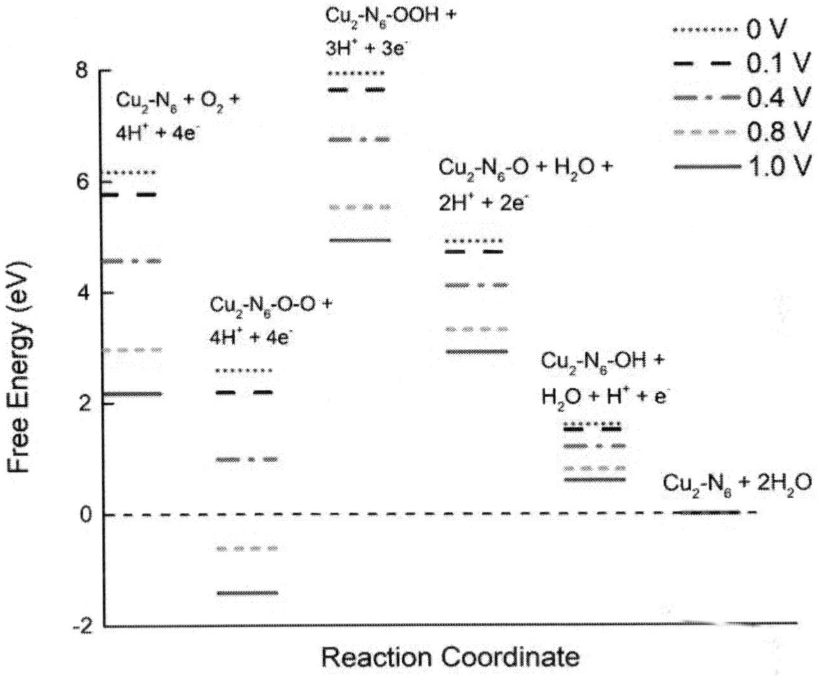


**Figure 3.8:** Free energy of possible ORR path for $Cu_2N_6$/Gr (01) defect structure with different potentials in an acidic medium.

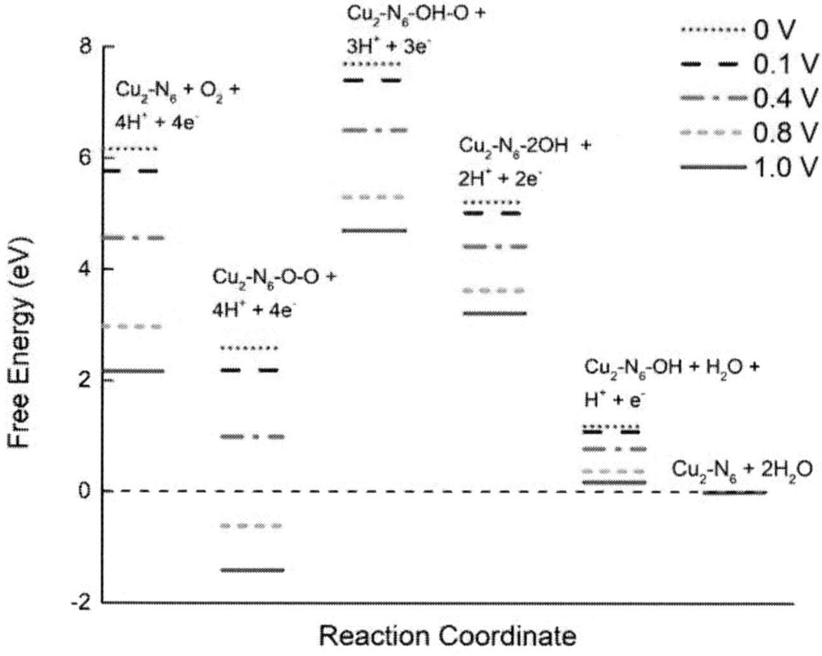


**Figure 3.9:** Free energy of possible ORR paths for $Cu_2$-$N_6$/Gr (02) defect structures with different potentials in acidic medium.

The $Cu_2$-$N_8$/Gr structure shows spontaneous reaction for all ORR steps for cell potential ranges from 0 to 0.28 V. With cell potential >0.28V, the fifth step of ORR becomes uphill, showing the *2OH → *OH + $H_2O$ reaction process is not spontaneous (**Figure 3.10**). Further increase in the cell potential caused an increase in the uphill processes. The Cu-$N_4$/Gr structure shows a spontaneous process for all ORR steps until the cell potential reaches 0.49 V. After the cell potential exceeds 0.49 V ($U > 0.49$ V), the ORR step of forming *O + $H_2O$ becomes uphill (**Figure 3.11**). Then, with the enhancement of the potential, the uphill process also increased. The maximum cell potentials for spontaneous reactions are shown to be 0.49 and 0.28 V for Cu-$N_4$/Gr and $Cu_2$-$N_8$/Gr, respectively. These results show that the open circuit voltage (OCV) is relatively better than the Pt catalyst cathode, which initially shows 0.9 V, as reported by Kim et. al. [40].

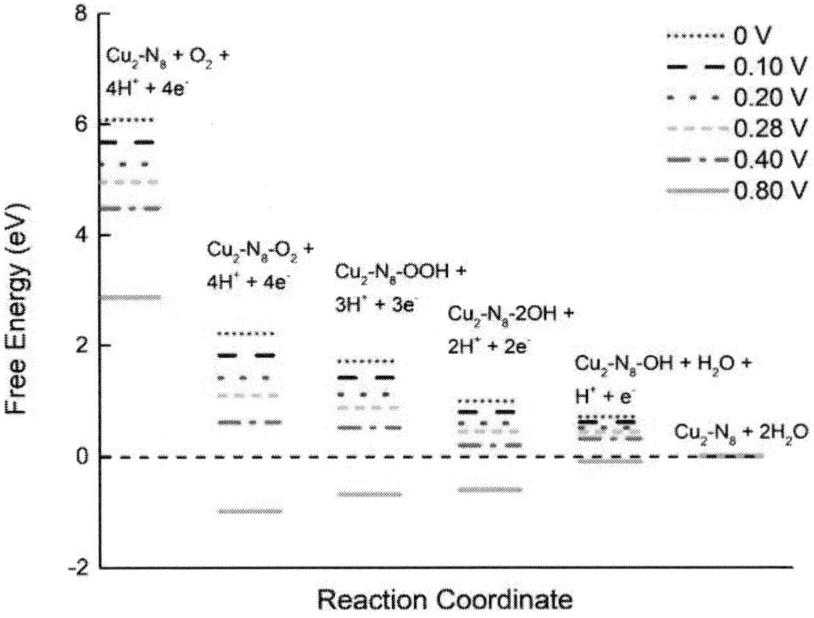


**Figure 3.10:** Free energy of possible ORR paths for $Cu_2$-$N_8$/Gr defect structures with different potentials in acidic medium.

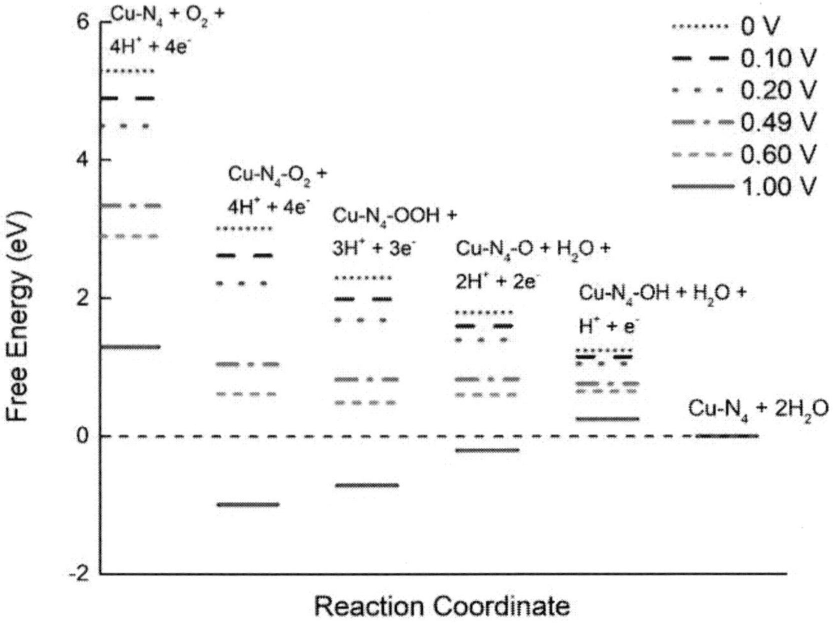


**Figure 3.11:** Free energy of possible ORR paths for Cu-$N_4$/Gr defect structures with different potentials in acidic medium.

A comparison of formation energies and chemical hardness indicates that the Cu-$N_4$/Gr structure has less probability of forming and is less stable than other structures. Therefore, generating 0.49 V as cell voltage is a low possibility. Furthermore, the Cu-$N_4$/Gr defect structure has a greater possibility of forming $H_2O_2$ as indicated in **Table 3.2**. The free energy diagram shows (

**Figure 3.12**) $H_2O_2$ formation kept spontaneously within the range of 0 - 0.49 V. After exceeding 0.49 V, the $H_2O_2$ formation step becomes uphill. Both the free energy and the maximum potential for the downhill process to form $H_2O_2$ and $H_2O$ from *OOH are the same. Furthermore, Mulliken charges of O atoms of *OOH are approximately equal (-0.312 e and -0.331 e, respectively), indicating that the final step of ORR of Cu-$N_4$/Gr is solely dependent on the binding of the H atom to the O atom of *OOH (**Figure 3.7 (b)**). If the H atom binds to the Cu-O side O atom of *OOH, it could generate $H_2O_2$, and binding to the other O atom would generate $2H_2O$ in the final step. Since the $H_2O_2$ formation path supplies only two electrons through an external path, from the anode to the cathode, could result in a decrease in current density.

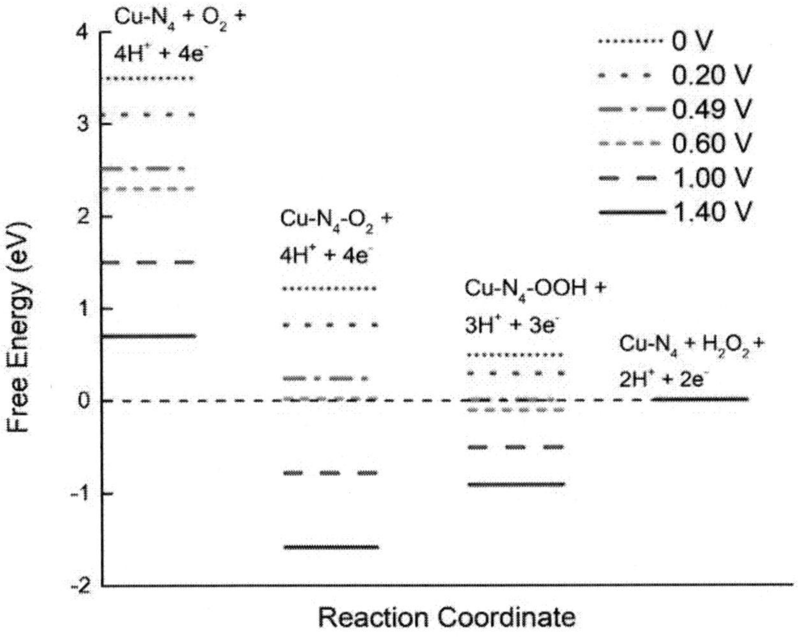


**Figure 3.12:** Free energy diagram of $H_2O_2$ generation in Cu-$N_4$/Gr defect structures with different potentials in acidic medium.

# CHAPTER 4

## CONCLUSION AND OUTLOOK

### 4.1 Conclusion

Computational calculations predicted the $Cu_2$-$N_8$/Gr and Cu-$N_4$/Gr motifs, showing promising spontaneous ORR with maximum cell potential of 0.28 and 0.49 V under standard conditions. The $Cu_2$-$N_6$/Gr motif is not favorable for spontaneous ORR. From the formation energy calculations, $Cu_2$-$N_8$/Gr, Cu-$N_4$/Gr, and $Cu_2$-$N_6$/Gr are stable for potential range (U) 0 - 5.87, 0 - 6.6, and 0 - 5.89 V, respectively. The maximum cell potentials of $Cu_2$-$N_8$/Gr and Cu-$N_4$/Gr are within the potential range of formation energy, showing structural stability. Based on the chemical hardness, chemical potential, electrophilicity, and Mulliken charge calculations, the $Cu_2$-$N_8$/Gr shows a higher stability and favors ORR than $Cu_2$-$N_6$/Gr. $Cu_2$-$N_6$/Gr shows more favor for ORR than Cu-$N_4$/Gr according to the binding energy values. The $Cu_2$-$N_6$/Gr motif is not recognized as a spontaneous catalyst by free energy calculations. Therefore, considering these three motifs, only $Cu_2$-$N_8$/Gr and Cu-$N_4$/Gr shows promising ORR ability. Consequently, the possibility of cell voltage solely depends on $Cu_2$-$N_8$/Gr could occur due to its high stability. According to DFT optimization, $H_2O_2$ is generated only from Cu-$N_4$/Gr. The other structures follow four-electron transfer pathways to generate high electron density. Cu-Nitrogen doped non-PGM catalyst with the considered motifs has a relatively high current density even with low cell voltage compared to Pt-like catalysts. The amount of $H_2O_2$ generated by Cu-$N_4$/Gr could be limited by its low stability. Bhatt et al have also shown that the Cu-$N_2$/Gr motif has a high probability of generating $H_2O_2$, but the -5.68 eV formation energy [21] indicates that it has lower stability than Cu-$N_4$/Gr. In this study, a novel $Cu_2$-$N_8$/Gr catalyst is a good candidate for fuel cell applications as it has good cell potential and non-$H_2O_2$ formation. Furthermore, the three structures presented in this study show a better $O_2$ binding energy than the Pt catalyst and PGM alloys [41].

### 4.2 Outlook

This study has several limitations that should be acknowledged. First, the 3-21G basis set used in this study can be considered small. However, it was selected based on Bhatt et al.'s study [21], which compared Cu-$N_2$/Gr results with the current study. Furthermore, this study did not include periodic boundary conditions or dispersion interactions, unlike other studies, such as Xiao et al.'s study[20] on Cu-$N_4$/Gr. However, these studies cannot be used as a basis for comparison with the current study due to their use of different DFT simulation software. The use of Gaussian09 instead of VASP-like simulation software was due to financial constraints. While VASP-like software is better suited for surface simulation and Periodic boundary condition (PBC) approach calculations, this study employed a single layer of graphene to examine the impact of defect structures on the surface of the catalyst in relation to ORR, similar to the approaches taken in Bhatt et al. [21] and Kattel et al. [16]. Future studies related to this research will employ dispersion correction with the PBC approach to provide a basis for comparison with the current study.

Few transitional metals like Mn, Fe, Co with nitrogen as a defect in graphene layer shown considerably good potential for low temperature fuel cell catalyst. Even with these metals, previous studies only consider on few defect structures like M-$N_2$/Gr, M-$N_4$/Gr where M is transitional metal. Also, almost every DFT study done in initial temperature (298.15 K).

Since the current study has shown the $Cu_2$-$N_8$/Gr motif has better stability and spontaneous ORR, future investigations with other transitional metals with the $M_2$-$N_8$/Gr motif at different temperatures (above 353.15 K) could be valuable. For this new study, adding dispersion correction will bring the results closer to realistic values. Additionally, DFT simulation on 3D bulk structure also could shed some light to new information on non-PGM catalysts.

# CHAPTER 5

## REFERENCE


1. *Fuel Cells*. [cited 2022 8th January 2022]; Available from: https://www.energy.gov/eere/fuelcells/fuel-cells.
2. Antolini, E., *Carbon supports for low-temperature fuel cell catalysts.* Appl. Catal. B: Environmental, 2009. **88**(1-2): p. 1-24.
3. Roland, U.B., Thomas Roessner, Frank, *On the nature of spilt-over hydrogen.* J. Mol. Catal. A: Chem, 1997. **127**(1-3): p. 61-84.
4. Conner Jr, W.C., G. Pajonk, and S. Teichner, *Spillover of Sorbed Species*, in *Advances in catalysis*. 1986, Elsevier. p. 1-79.
5. Lin, R., et al., *Rapid microwave-assisted solvothermal synthesis of shape-controlled Pt-Ni alloy nanoparticles for PEMFC.* Electrochim. Acta, 2018. **283**: p. 764-771.
6. Reyimjan A. Sidik, A.B.A., *O2 Reduction on Graphite and Nitrogen-Doped Graphite Experiment and Theory.* J. Phys. Chem, 2006. **110**: p. 1787-1793.
7. Lee, K.R., et al., *Electrochemical oxygen reduction on nitrogen doped graphene sheets in acid media.* Electrochem. commun, 2010. **12**(8): p. 1052-1055.
8. Zhang, L. and Z. Xia, *Mechanisms of Oxygen Reduction Reaction on Nitrogen-Doped Graphene for Fuel Cells.* J. Phys. Chem C, 2011. **115**(22): p. 11170-11176.
9. Rao, C.V., C.R. Cabrera, and Y. Ishikawa, *In Search of the Active Site in Nitrogen-Doped Carbon Nanotube Electrodes for the Oxygen Reduction Reaction.* J. Phys. Chem Letters, 2010. **1**(18): p. 2622-2627.
10. Lin, R., et al., *High durability of Pt-Ni-Ir/C ternary catalyst of PEMFC by stepwise reduction synthesis.* Electrochim. Acta, 2020. **330**.
11. Beermann, V., et al., *Real-time imaging of activation and degradation of carbon supported octahedral Pt–Ni alloy fuel cell catalysts at the nanoscale using in situ electrochemical liquid cell STEM.* Energy Environ. Sci. , 2019. **12**(8): p. 2476-2485.
12. Alexei L.N. Pinheiro, A.O.-N., Elki C. de Souza,, V.A.P. Joelma Perez, Edson A. Ticianelli, and a.E.R. Gonzalez, *Electrocatalysis-on-Noble-Metal-and-Noble-Metal-Alloys-Dispersed-on-High-Surface-Area-Carbon.* J. New Mater. Electrochem. Sys, 2003. **6**: p. 1-8.
13. Orellana, W., *Catalytic Properties of Transition Metal–N4 Moieties in Graphene for the Oxygen Reduction Reaction: Evidence of Spin-Dependent Mechanisms.* J. Phys. Chem C, 2013. **117**(19): p. 9812-9818.
14. Khotseng, L., *Oxygen Reduction Reaction*, in *Electrocatalysts for Fuel Cells and Hydrogen Evolution - Theory to Design*. 2018. p. 27.
15. Liu, K., et al., *Mn- and N- doped carbon as promising catalysts for oxygen reduction reaction: Theoretical prediction and experimental validation.* Appl. Catal. B: Environ, 2019. **243**: p. 195-203.
16. Kattel, S., P. Atanassov, and B. Kiefer, *Catalytic activity of Co-N(x)/C electrocatalysts for oxygen reduction reaction: a density functional theory study.* Phys Chem Chem Phys, 2013. **15**(1): p. 148-53.
17. Masa, J., et al., *Metal-free catalysts for oxygen reduction in alkaline electrolytes: Influence of the presence of Co, Fe, Mn and Ni inclusions.* Electrochim. Acta, 2014. **128**: p. 271-278.
18. Na Yang, I.P., Li Li, Jing Li, Qiang Liao, Minhua Shao, Zidong Wei, *Theoretically probing the possible degradation mechanisms of an FeNC catalyst during the oxygen reduction reaction.* Chem.Sci, 2021: p. 12476-12484.
19. Xie, X., et al., *Performance enhancement and degradation mechanism identification of a single-atom Co–N–C catalyst for proton exchange membrane fuel cells.* Nat. Catal, 2020. **3**(12): p. 1044-1054.
20. Xiao, Y. and W. Zhang, *DFT analysis elementary reaction steps of catalytic activity for ORR on metal-, nitrogen- co-doped graphite embedded structure.* SN Appl. Sci, 2020. **2**(2): p. 1-11.
21. Bhatt, M.D., G. Lee, and J.S. Lee, *Density Functional Theory (DFT) Calculations for Oxygen Reduction Reaction Mechanisms on Metal-, Nitrogen- co-doped Graphene (M-N2-G (M = Ti, Cu, Mo, Nb and Ru)) Electrocatalysts.* Electrochim. Acta, 2017. **228**: p. 619-627.

22. Chen, C., et al., *Dual-metal single-atomic catalyst: The challenge in synthesis, characterization, and mechanistic investigation for electrocatalysis.* SmartMat, 2022: p. 1–32.
23. Zhang, N., et al., *High-Density Planar-like Fe2N6 Structure Catalyzes Efficient Oxygen Reduction.* Matter, 2020. **3**(2): p. 509-521.
24. J. K. Nørskov, J.R., A. Logadottir, and L. Lindqvist, *Origin of the Overpotential for Oxygen Reduction at a Fuel-Cell Cathode.* J. Phys. Chem B, 2004. **108**: p. 17886-17892.
25. Calle-Vallejo, F., J.I. Martinez, and J. Rossmeisl, *Density functional studies of functionalized graphitic materials with late transition metals for Oxygen Reduction Reactions.* Phys Chem Chem Phys, 2011. **13**(34): p. 15639-43.
26. Rongrong Chen, H.L., Deryn Chu and Guofeng Wang, *Unraveling Oxygen Reduction Reaction Mechanisms on Carbon-Supported Fe-Phthalocyanine and Co-Phthalocyanine Catalysts in Alkaline Solutions.* J. Phys. Chem., 2009(113): p. 20689-20697.
27. Stephen Walch, A.D., Masoud Aryanpour, and Heinz Pitsch, *Mechanism of Molecular Oxygen Reduction at the Cathode of a PEM Fuel Cell Non-Electrochemical Reactions on Catalytic Pt Particles.* J. Phys. Chem, 2008. **112**: p. 8464–8475.
28. Yu, L., et al., *Oxygen reduction reaction mechanism on nitrogen-doped graphene: A density functional theory study.* J. Catal, 2011. **282**(1): p. 183-190.
29. Syaahiran, M.A., et al., *Theoretical Study of CO Adsorption Interactions with Cr-Doped Tungsten Oxide/Graphene Composites for Gas Sensor Application.* ACS Omega, 2022. **7**(1): p. 528-539.
30. Glossman-Mitnik, D., *Computational Study of the Chemical Reactivity Properties of the Rhodamine B Molecule.* Procedia Comput. Sci, 2013. **18**: p. 816-825.
31. Chattaraj, P.K. and S. Giri, *Electrophilicity index within a conceptual DFT framework.* Annu. Rep. Section C: Phys. Chem, 2009. **105**.
32. Parr, R.G., L.v. Szentpaly, and S. Liu, *Electrophilicity Index.* J. Am. Chem. Soc., 1999. **121**: p. 1922-1924.
33. Syaahiran, A., et al., *A theoretical insight of Cr dopant in tungsten oxide for gas sensor application.* Mater. Today Commun, 2021. **28**: p. 102508.
34. Sha, Y., et al., *Theoretical Study of Solvent Effects on the Platinum-Catalyzed Oxygen Reduction Reaction.* J. Phys. Chem Letters, 2010. **1**(5): p. 856-861.
35. M. Isaacs, J.C.C., M.J. Aguirre, G. Estiu´, F. Caruso, G. Ferraudi, and J. Costamagna, *Electrocatalytic reduction of CO2 by aza-macrocyclic complexes of Ni(II), Co(II), and Cu(II). Theoretical contribution to probable mechanisms.* Inorg. Chim. Acta, 2002. **339**: p. 224-232.
36. Nallathambi, V., et al., *Development of high performance carbon composite catalyst for oxygen reduction reaction in PEM Proton Exchange Membrane fuel cells.* J. Power Sources, 2008. **183**(1): p. 34-42.
37. Satoru Hommura, K.K., Tetsuji Shimohira, and and Y. Teraoka, *Development of a Method for Clarifying the Perfluorosulfonated Membrane Degradation Mechanism in a Fuel Cell Environment.* J. Electrochem. Soc., 2008. **155**(1): p. A29-A33.
38. Cheng Chen, T.F.F., *H2O2 Formation under Fuel-cell Conditions.* J. Electrochem. Soc., 2007. **11**(1): p. 1127-1137.
39. Chai, G.-L., et al., *Active sites engineering leads to exceptional ORR and OER bifunctionality in P,N Co-doped graphene frameworks.* Energy Environ. Sci. , 2017. **10**(5): p. 1186-1195.
40. Kim, J., J. Lee, and Y. Tak, *Relationship between carbon corrosion and positive electrode potential in a proton-exchange membrane fuel cell during start/stop operation.* J. Power Sources, 2009. **192**(2): p. 674-678.
41. Qi, X., et al., *DFT study on ORR catalyzed by bimetallic Pt-skin metals over substrates of Ir, Pd and Au.* Nano Mater. Sci, 2021.

# APPENDICES

**Appendix 1:** Gaussian DFT calculated energy data for defect and pristine graphene structure (without hydrogen atoms).

| STRUCTURE | ENERGY OF DEFECT STRUCTURE / Ha | PRISTINE GRAPHENE STRUCTURE / Ha |
|---|---|---|
| $Cu\text{-}N_2/Gr$ | -3410.97114 | -1821.05254 |
| $Cu\text{-}N_4/Gr$ | -3444.128924 | -1821.05254 |
| $Cu_2\text{-}N_6/Gr$ | -5034.085500 | -1821.05254 |
| $Cu_2\text{-}N_8/Gr$ | -5067.09674 | -1821.05254 |

**Appendix 2:** Gaussian data for binding energy calculations. X, $E_{(M\text{-}N+X)}$, $E_{(M\text{-}N)}$, $E_{(X)}$ are intermediate, optimized energy of intermediate attach to defect structure, optimized energy of defect structure and optimized energy of intermediate respectively.

| STRUCTURE | X | $E_{(M\text{-}N+X)}$ / Ha | $E_{(M\text{-}N)}$ / Ha | $E_{(X)}$ / Ha |
|---|---|---|---|---|
| | $H_2O$ | -3495.938405z | -3419.922739 | -75.973965 |
| | O | -3494.685444 | -3419.922739 | -74.554918 |
| $Cu\text{-}N_2$ | $O_2$ | -3569.433891 | -3419.922739 | -149.419122 |
| | OH | -3495.305925 | -3419.922739 | -75.303138 |
| | OOH | -3570.042101 | -3419.922739 | -150.051099 |
| | $H_2O_2$ | -3570.659309 | -3419.922739 | -150.699782 |
| | $H_2O$ | -3529.099665 | -3453.080522 | -75.973965 |
| | O | -3527.824567 | -3453.080522 | -74.554918 |
| | O2 | -3602.601097 | -3453.080522 | -149.419122 |
| $Cu\text{-}N_4$ | OH | -3528.456971 | -3453.080522 | -75.303138 |
| | OOH | -3603.194155 | -3453.080522 | -150.051099 |
| | $H_2O_2$ | -3603.81402 | -3453.080522 | -150.699782 |
| | $H_2O$ | -5119.058805 | -5043.037098 | -75.973965 |
| | O | -5117.868674 | -5043.037098 | -74.554918 |
| $Cu_2\text{-}N_6$ (01) | O+O | -5192.586386 | -5043.037098 | -149.419122 |
| | OH | -5118.528562 | -5043.037098 | -75.303138 |
| | OOH | -5193.240117 | -5043.037098 | -150.051099 |
| | $OH+OH(H_2O_2)$ | -5193.843348 | -5043.037098 | -150.606276 |
| | $H_2O$ | -5119.058805 | -5043.037098 | -75.973965 |
| | O + OH | -5193.245112 | -5043.037098 | -149.858056 |
| $Cu_2\text{-}N_6$ (02) | O+O | -5192.595087 | -5043.037098 | -149.419122 |
| | OH + OH | -5193.882811 | -5043.037098 | -150.606276 |
| | OH | -5118.528815 | -5043.037098 | -75.303138 |
| | $H_2O_2$ | - | | |
| | $H_2O$ | -5152.058235 | -5076.048338 | -75.973965 |
| | O | - | - | - |
| | $O_2$ | -5225.569916 | -5076.048338 | -149.419122 |
| $Cu_2\text{-}N_8$ | OH + OH | -5226.850944 | -5076.048338 | -150.606276 |
| | OOH | -5226.219613 | -5076.048338 | -150.051099 |
| | OH | -5151.491395 | -5076.048338 | -75.303138 |
| | $H_2O_2$ | - | - | - |

**Appendix 3:** Gaussian data for free energy calculation. EE. TE, S, ZPE are electronic energy, thermal energy, entropy, and correction of zero-point energy respectively.

| ATOM / INTERMEDIATE | DATA | VALUE |
|---|---|---|
| H | EE /Ha | -0.497311 |
| | TE /Kcal $mol^{-1}$ | -0.494951 |
| | S /cal $k^{-1}$ $mol^{-1}$ | 27.392 |
| | ZPE /Ha | 0 |
| $O_2$ | EE /Ha | -149.419122 |
| | TE /Kcal $mol^{-1}$ | -149.412686 |
| | S /cal $k^{-1}$ $mol^{-1}$ | 47.126 |
| | ZPE /Ha | 0.003123 |
| OOH | EE /Ha. | -150.051099 |
| | TE /Kcal $mol^{-1}$ | 10.158 |
| | S /cal $k^{-1}$ $mol^{-1}$ | 55.096 |
| | ZPE /Ha | 0.013053 |
| O | EE /Ha | -74.554918 |
| | TE /Kcal $mol^{-1}$ | 0.889 |
| | S /cal $k^{-1}$ $mol^{-1}$ | 38.494 |
| | ZPE /Ha | 0 |
| OH | EE /Ha | -75.303138 |
| | TE /Kcal $mol^{-1}$ | 6.129 |
| | S /cal $k^{-1}$ $mol^{-1}$ | 42.777 |
| | ZPE /Ha | 0.007406 |
| $H_2O_2$ | EE /Ha | -150.699782 |
| | TE /Kcal $mol^{-1}$ | 17.477 |
| | S /cal $k^{-1}$ $mol^{-1}$ | 58.423 |
| | ZPE /Ha | 0.024116 |
| $H_2O$ | EE /Ha. | -75.973965 |
| | TE /Kcal $mol^{-1}$ | -75.950443 |
| | S /cal $k^{-1}$ $mol^{-1}$ | 45.304 |
| | ZPE /Ha | 0.019743 |
| $H_2$ | EE /Ha | -1.17061101 |
| | TE /Kcal $mol^{-1}$ | - |
| | S /cal $k^{-1}$ $mol^{-1}$ | 31.151 |
| | ZPE /Ha | 0.010184 |

**Appendix 4:** $Cu-N_4/Gr$ structure - Gaussian data for the graph of free energy Vs. reaction coordinates with different potentials.

| REACTION COORDINATES | FREE ENERGY FOR DIFFERENT POTENTIAL (U) / eV | | | | | |
|---|---|---|---|---|---|---|
| | 0.00 V | 0.10 V | 0.20 V | 0.49 V | 0.60 V | 1.00 V |
| $Cu-N_4 + O_2 + 4H + 4e$ | 5.30 | 4.90 | 4.50 | 3.34 | 2.90 | 1.30 |
| $Cu-N_4-O_2 + 4H + 4e$ | 3.01 | 2.61 | 2.21 | 1.05 | 0.61 | -0.99 |
| $Cu-N_4-OOH + 3H + 3e$ | 2.29 | 1.99 | 1.69 | 0.82 | 0.49 | -0.71 |
| $Cu-N_4-O + 2H + H_2O + 2e$ | 1.80 | 1.60 | 1.40 | 0.82 | 0.60 | -0.20 |
| $Cu-N_4-OH + H + H_2O + e$ | 1.25 | 1.15 | 1.05 | 0.76 | 0.65 | 0.25 |
| $Cu-N_4 + H_2O + H_2O$ | 0.00 | 0.00 | 0.00 | 0.00 | 0.00 | 0.00 |

**Appendix 5:** $Cu_2-N_6/Gr$ (01) structure - Gaussian data for the graph of free energy Vs. reaction coordinates with different potentials.

| REACTION COORDINATES | FREE ENERGY FOR DIFFERENT POTENTIAL (U) / eV | | | | |
|---|---|---|---|---|---|
| | 0 V | 0.1 V | 0.40 V | 0.80 V | 1.00 V |
| $Cu_2-N_6 + O_2 + 4H + 4e$ | 6.17 | 5.77 | 4.57 | 2.97 | 2.17 |
| $Cu_2-N_6-O-O + 4H + 4e$ | 2.59 | 2.19 | 0.99 | -0.61 | -1.41 |
| $Cu_2-N_6-OOH + 3H + 3e$ | 7.93 | 7.63 | 6.73 | 5.53 | 4.93 |
| $Cu_2-N_6-O + 2H + H_2O + 2e$ | 4.91 | 4.71 | 4.11 | 3.31 | 2.91 |
| $Cu_2-N_6-OH + H + H_2O + e$ | 1.60 | 1.50 | 1.20 | 0.80 | 0.60 |
| $Cu_2-N_6 + H_2O + H_2O$ | 0.00 | 0.00 | 0.00 | 0.00 | 0.00 |

**Appendix 6:** $Cu_2-N_6/Gr$ (02) structure - Gaussian data for the graph of free energy Vs. reaction coordinates with different potentials.

| REACTION COORDINATES | FREE ENERGY FOR DIFFERENT POTENTIAL (U) / eV | | | | |
|---|---|---|---|---|---|
| | 0 V | 0.1 V | 0.40 V | 0.80 V | 1.00 V |
| $Cu_2-N_6 + O_2 + 4H + 4e$ | 6.17 | 5.77 | 4.57 | 2.97 | 2.17 |
| $Cu_2-N_6-O-O + 4H + 4e$ | 2.60 | 2.20 | 1.00 | -0.60 | -1.40 |
| $Cu_2-N_6-OH-O + 3H + 3e$ | 7.71 | 7.41 | 6.51 | 5.31 | 4.71 |
| $Cu_2-N_6-2OH + 2H + 2e$ | 5.22 | 5.02 | 4.42 | 3.62 | 3.22 |
| $Cu_2-N_6-OH + H_2O + H + e$ | 1.18 | 1.08 | 0.78 | 0.38 | 0.18 |
| $Cu_2-N_6 + H_2O + H_2O$ | 0.00 | 0.00 | 0.00 | 0.00 | 0.00 |

**Appendix 7:** $Cu_2$-$N_8$/Gr structure - Gaussian data for the graph of free energy Vs. reaction coordinates with different potentials.

| REACTION COORDINATES | FREE ENERGY FOR DIFFERENT POTENTIAL (U) / eV | | | | | |
|---|---|---|---|---|---|---|
| | 0.00 V | 0.10 V | 0.20 V | 0.28 V | 0.40 V | 0.80 V |
| $Cu_2$-$N_8$ + $O_2$ + 4H +4e | 6.08 | 5.68 | 5.28 | 4.96 | 4.48 | 2.88 |
| $Cu_2$-$N_8$-$O_2$ + 4H +4e | 2.22 | 1.82 | 1.42 | 1.10 | 0.62 | -0.98 |
| $Cu_2$-$N_8$-OOH + 3H +3e | 1.71 | 1.41 | 1.11 | 0.87 | 0.51 | -0.69 |
| $Cu_2$-$N_8$-2OH + 2H +2e | 0.99 | 0.79 | 0.59 | 0.43 | 0.19 | -0.61 |
| $Cu_2$-$N_8$-OH + $H_2O$ + H +e | 0.71 | 0.61 | 0.51 | 0.43 | 0.31 | -0.09 |
| $Cu_2$-$N_8$ + $H_2O$ + $H_2O$ | 0.00 | 0.00 | 0.00 | 0.00 | 0.00 | 0.00 |

**Appendix 8:** The Gaussian data of the graph of free energy Vs. reaction coordinates with different potentials for $H_2O_2$ form by Cu-$N_4$/Gr structure.

| REACTION COORDINATES | FREE ENERGY FOR DIFFERENT POTENTIAL (U) / eV | | | | | |
|---|---|---|---|---|---|---|
| | 0.00 | 0.20 | 0.49 | 0.60 | 1.00 | 1.40 |
| Cu-$N_4$ + $O_2$ + 4H + 4e | 3.5014 | 3.1014 | 2.52 | 2.30 | 1.50 | 0.70 |
| Cu-$N_4$-$O_2$ + 4H + 4e | 1.2135 | 0.8135 | 0.23 | 0.01 | -0.79 | -1.59 |
| Cu-$N_4$-OOH + 3H +3e | 0.4924 | 0.2924 | 0.00 | -0.11 | -0.51 | -0.91 |
| Cu-$N_4$ + $H_2O_2$ + 2H + 2e | 0.0000 | 0.0000 | 0.00 | 0.00 | 0.00 | 0.00 |

**Appendix 9:** Ionization potential (I), electron affinity (A), Energy gap ($E_g$), chemical hardness(η), chemical potential (μ), and electrophilicity index (ω) values of intermediates and defect structures (Gaussian data).

| STRUCTURE | INTERMEDIATES | HOMO /eV | LUMO /eV | I /eV | A /eV | Eg /eV | η /eV | μ /eV | ω /eV |
|---|---|---|---|---|---|---|---|---|---|
| | $Cu\text{-}N_4$ | -0.14113 | -0.11165 | 0.14113 | 0.11165 | 0.03 | 0.03 | -0.13 | 0.27 |
| | $H_2O$ | -0.13574 | -0.09415 | 0.13574 | 0.09415 | 0.04 | 0.04 | -0.11 | 0.16 |
| | O | -0.14872 | -0.10446 | 0.14872 | 0.10446 | 0.04 | 0.04 | -0.13 | 0.18 |
| **$Cu\text{-}N_4$** | $O_2$ | -0.14751 | -0.10534 | 0.14761 | 0.10534 | 0.04 | 0.04 | -0.13 | 0.19 |
| | OH | -0.14593 | -0.11089 | 0.14593 | 0.11089 | 0.04 | 0.04 | -0.13 | 0.24 |
| | OOH | -0.14587 | -0.11082 | 0.14587 | 0.11082 | 0.04 | 0.04 | -0.13 | 0.23 |
| | $H_2O_2$ | -0.1331 | -0.10015 | 0.1331 | 0.10015 | 0.03 | 0.03 | -0.12 | 0.21 |
| | $Cu_2\text{-}N_6$ | -0.14942 | -0.11281 | 0.14942 | 0.11281 | 0.04 | 0.04 | -0.13 | 0.23 |
| | $H_2O$ | -0.14068 | -0.10669 | 0.14068 | 0.10669 | 0.03 | 0.03 | -0.12 | 0.23 |
| | O | -0.15155 | -0.11166 | 0.15155 | 0.11166 | 0.04 | 0.04 | -0.13 | 0.22 |
| **$Cu_2\text{-}N_6$ (01)** | O+O | -0.1545 | -0.12477 | 0.1545 | 0.12477 | 0.03 | 0.03 | -0.14 | 0.33 |
| | OH | -0.15354 | -0.08538 | 0.15354 | 0.08538 | 0.07 | 0.07 | -0.12 | 0.10 |
| | OOH | -0.15362 | -0.08827 | 0.15362 | 0.08827 | 0.07 | 0.07 | -0.12 | 0.11 |
| | OH+OH($H_2O_2$) | -0.15761 | -0.12435 | 0.15761 | 0.12435 | 0.03 | 0.03 | -0.14 | 0.30 |
| | $H_2O$ | -0.14069 | -0.10668 | 0.14069 | 0.10668 | 0.03 | 0.03 | -0.12 | 0.22 |
| | O + OH | -0.16236 | -0.08468 | 0.16236 | 0.08468 | 0.08 | 0.08 | -0.12 | 0.10 |
| **$Cu_2\text{-}N_6$ (02)** | O+O | -0.15245 | -0.12502 | 0.15245 | 0.12502 | 0.03 | 0.03 | -0.14 | 0.35 |
| | OH + OH | -0.15222 | -0.1188 | 0.15222 | 0.1188 | 0.03 | 0.03 | -0.14 | 0.27 |
| | OH | -0.14436 | -0.11488 | 0.14436 | 0.11488 | 0.03 | 0.03 | -0.13 | 0.28 |
| | $Cu_2\text{-}N_8$ | -0.13986 | -0.09484 | 0.13986 | 0.09484 | 0.05 | 0.05 | -0.12 | 0.15 |
| | $H_2O$ | -0.13748 | -0.08922 | 0.13748 | 0.08922 | 0.05 | 0.05 | -0.11 | 0.13 |
| | $O_2$ | -0.14785 | -0.10563 | 0.14785 | 0.10563 | 0.04 | 0.04 | -0.13 | 0.19 |
| **$Cu_2\text{-}N_8$** | OH + OH | -0.14824 | -0.10591 | 0.14824 | 0.10591 | 0.04 | 0.04 | -0.13 | 0.19 |
| | OOH | -0.13524 | -0.10302 | 0.13524 | 0.10302 | 0.03 | 0.03 | -0.12 | 0.22 |
| | OH | -0.13736 | -0.08121 | 0.13736 | 0.08121 | 0.06 | 0.06 | -0.11 | 0.11 |

**Appendix 10:** Quantum Espresso DFT calculated energy data for defect and pristine graphene structure (without hydrogen atoms).

| STRUCTURE | ENERGY OF DEFECT STRUCTURE / Ry | PRISTINE GRAPHENE STRUCTURE / Ry |
|---|---|---|
| $Cu-N_4/Gr$ | -1137.315755 | -922.188068 |
| $Cu_2-N_6/Gr$ | -1332.986477 | -922.188068 |
| $Cu_2-N_8/Gr$ | -1352.313347 | -922.188068 |

**Appendix 11:** Quantum Espresso data for binding energy calculations. X, $E_{(M\text{-}N + X)}$, $E_{(M\text{-}N)}$, $E_{(X)}$ are intermediate, optimized energy of intermediate attach to defect structure, optimized energy of defect structure and optimized energy of intermediate, respectively.

| STRUCTURE | X | $E_{(M\text{-}N + X)}$ / Ry | $E_{(M\text{-}N)}$ / Ry | $E_{(X)}$ / Ry |
|---|---|---|---|---|
| $Cu-N_4$ | $H_2O$ | -1181.358022 | -1137.315755 | -44.03856731 |
| | O | -1178.746033 | -1137.315755 | -41.17117386 |
| | O2 | -1220.32342 | -1137.315755 | -82.96571523 |
| | OH | -1180.073245 | -1137.315755 | -42.60309724 |
| | OOH | -1221.557401 | -1137.315755 | -84.17388637 |
| | $H_2O_2$ | -1222.802415 | -1137.315755 | -85.4838626 |

Publisher: Eliva Press Global Ltd

Email: info@elivapress.com